Abraão Martins Terceiro

SAO Letters: An analysis in the context of environmental justice

Abraão Martins Terceiro

SAO Letters: An analysis in the context of environmental justice

Case study in Tramandaí and Imbé - RS, Brazil

ScienciaScripts

Imprint
Any brand names and product names mentioned in this book are subject to trademark, brand or patent protection and are trademarks or registered trademarks of their respective holders. The use of brand names, product names, common names, trade names, product descriptions etc. even without a particular marking in this work is in no way to be construed to mean that such names may be regarded as unrestricted in respect of trademark and brand protection legislation and could thus be used by anyone.

Cover image: www.ingimage.com

This book is a translation from the original published under ISBN 978-3-330-76504-7.

Publisher:
Sciencia Scripts
is a trademark of
Dodo Books Indian Ocean Ltd. and OmniScriptum S.R.L publishing group

120 High Road, East Finchley, London, N2 9ED, United Kingdom
Str. Armeneasca 28/1, office 1, Chisinau MD-2012, Republic of Moldova, Europe
Managing Directors: Ieva Konstantinova, Victoria Ursu
info@omniscriptum.com

Printed at: see last page
ISBN: 978-620-8-51615-4

SUMMARY

The research addresses the importance of Environmental Sensitivity Letters to Oil Spills in the context of environmental management, with the aim of analysing this instrument from the perspective of Environmental Justice. To develop the work, a case study was carried out in the municipalities of Tramandaí and Imbé in the state of Rio Grande do Sul - Brazil, looking at the problems faced by artisanal fishermen in relation to oil activity, in general, and in relation to an accident involving an oil spill in January 2012, specifically. In this way, a study was set up with the aim of proposing recommendations that would protect families who depend on the coastal environment for their artisanal fishing. In order to acquire the data, four research procedures were carried out (field observation, field diary, interviews and workshops). The main result was the creation of a mental map, which showed the distribution of artisanal fishing in the study area and, finally, based on the results, led to a conclusion as to the fragility of fishing activity and its social reproduction in the event of oil spill accidents within the context of the SAO Charts.

Key words: SAO charters; environmental justice; vulnerability; artisanal fishing.

I dedicate this work to my mother and to the artisanal fishermen. May it be one more step towards recognising this admirable class.

SUMMARY

"Go quietly, fisherman, so as not to call the souls

If you hear the cry of the procellaria, come back, fisherman!

If you hear the bell of the Witch's Lighthouse, come back, fisherman!

If you hear the cry of the suicidal woman from the power station, come back, fisherman[1]"

The fisherman (Vinícius de Morais).

CHAPTER 1

INTRODUCTION

1.1. PRESENTATION

The flow of activities linked to the oil industry in Brazil is mostly linked to the coastal and maritime zone, through the transit of vessels, offshore platform installations, port facilities, oil transport pipelines and oil terminals. Despite the mastery of advanced techniques for transporting oil and oil products in the modes currently used (pipeline, rail, sea and road), the adverse conditions (human, environmental and material) faced by this system lead to accidents (MARTINS et *al.,* 2013).

According to Seifert Jr (2013), the impacts of this activity can occur at all stages (from exploration to refining), but it is at the transport stage that the worst damage can occur, with large-scale leaks from pipelines and oil tankers (Adame & Gambini, 2007). In addition, leaks can occur at the time of exploitation. In this way, we find a great danger to the marine and coastal environment.

The city of Tramandaí, located in the state of Rio Grande do Sul, is a coastal municipality that has oil flow activity in the region. In January 2012, an oil spill accident occurred in this municipality, which affected a large stretch of beach, causing a great deal of damage to the environment and society, both in the town itself and in the neighbouring municipality of Imbé.

In the event of accidents, Brazilian legislation provides for emergency plans so that the damage caused can be minimised and controlled in the shortest possible time. The Environmental Sensitivity to Oil Chart (SAO Chart) is a tool that aids decisions in this area (MARTINS et *al.,* 2013).

SAO charts are a management tool designed to deal with three types of situations involving oil-related activities: strategic planning, which is represented by the strategic sensitivity chart on a more macro scale in relation to the mapped region, covering the entire maritime sedimentary basin. The environmental licensing of oil activities, which can be represented by the use of tactical maps, and in the case of emergency and contingency measures for environmental accidents the use of operational maps.

Oil spills have caused recurrent and serious damage to coastal ecosystems in various regions of the world. When oil spills reach the coastal zone, they cause serious damage to wildlife, natural environments and socio-economic activities (WIECZORECK et *al.,* 2007).

With this scenario of impacts, it can be seen that conflicts over the use of space, such as the relationship with fishing, which is recurrent in Brazil (Walter & Anello, 2012), occur as a consequence of oil activity.

Studies by Serrão et *al.* (2009) and Walter & Anello (2012) show that the concentration and development of the oil industry's activities in the coastal area compete directly with the activities carried out by artisanal fishermen, among other social groups who make direct use of the space to carry out their occupations and means of work, and that this scenario is conducive to the occurrence of environmental conflicts, whether over the use of space or access to natural resources.

According to Cotrim & Miguel (2007), in the municipalities of Tramandaí and Imbé, artisanal fishing is considered to be an ancient activity that is extremely important for the region's economy. It involves a large number of people and families and is distributed in various areas throughout the municipalities.

The purpose of this study is to evaluate the methodology of environmental sensitivity charts for oil spills from the perspective of environmental justice.

In the literature that expresses the term Environmental Justice, it is possible to observe that various authors such as Bulard (1990); Acselrad (2004); Herculano & Pacheco (2006) defend the idea that no social group should disproportionately bear the impacts suffered by industrial and economic activities, due to their condition of social vulnerability.

Claims for environmental justice are relatively recent in the history of contemporary societies. It was only from the 1980s onwards that a form of discrimination produced by concentrating the consequences of natural degradation processes on certain social groups became notorious. Members of the social movements fighting for environmental justice thus began to denounce this discriminatory treatment, linking the need to maintain human and fundamental rights to the issue of environmental protection (BAGGIO, 2014).

In order to establish a connection between environmental justice and the context of environmental sensitivity maps to oil spills, the municipalities of Tramandaí and Imbé were chosen for this research, with the aim of carrying out a case study, due to the strategic consideration of the location of the region at the juncture between the existence of oil activity and the importance of artisanal fishing for the locality.

1.2. RESEARCH QUESTION

Do SAO charts, as a decision-making tool in the event of an oil spill, ensure the environmental integrity necessary for the social reproduction of artisanal fishermen?

1.3. OBJECTIVES

1.3.1 GENERAL OBJECTIVE

This paper focuses on the evaluation of environmental sensitivity charts for oil spills from the perspective of environmental justice.

1.3.2 SPECIFIC OBJECTIVES

The following specific objectives are proposed for the development of the work:

- Identify the main socio-environmental problems caused by oil spill accidents;

- To evaluate the methodology suggested and applied by the Ministry of the Environment used in the socio-economic characterisation included in the context of the SAO letters.

- Carry out a critical analysis based on the principles of environmental justice and through a case study in Tramandaí and Imbé/RS;

CHAPTER 2

STUDY AREA

Tramandaí and Imbé are Brazilian municipalities located on the northern coast of the state of Rio Grande do Sul. According to the 2010 IBGE census, Tramandaí has a population of 41,585 inhabitants and an area of 144.408 km^2 and the municipality of Imbé has 17,670 inhabitants and an area of 39.395 km2. The towns are neighbours and lie at latitude 29° 59' 05" south and longitude 50° 08' 01" west, 118 km from Porto Alegre, the capital of the state of Rio Grande do Sul (IBGE, 2010).

Figure 1 - Location of the study area.

Source: Google map data - 2014.

The municipalities of Tramandaí and Imbé were chosen for the research due to the fact that they are located in an area covered by the operational SAO chart of the Pelotas maritime basin. The region has activities linked to oil handling, with the presence of the water terminal and its monobuoys, linked to the Almirante Soares Dutra land terminal (TEDUT), along with the intense flow of large vessels travelling through the area, conflicting with fishing activities in the area.

The two municipalities are directly connected to the sea, surrounded by areas of intense water activity, with ecosystems of considerable ecological importance, featuring several areas of marshes, lagoons and rivers.

The existence of a large water supply in the municipalities has helped in the emergence of numerous activities linked to the environment, among which we can highlight artisanal fishing as one of the most practised and most observed in terms of the contact between human beings and the aquatic environment.

The Tramandaí Lagoon communicates with the sea and is the outlet for the drains of the entire Tramandaí river basin. This basin covers a total area in the region of 2,700 km^2 and its coastal strip is approximately 115 km long. It stretches from the sources of the Maquiné and Três Forquilhas rivers to the west, north to the Itapeva lagoon and south to the Cerquinha lagoon. This point of contact between the waters of the lagoons and the sea creates an estuarine zone (TRAMANDAÍ RIVER BATCHES COMMITTEE, 2005). As such, this region presents favourable conditions for the development of various species of fish, making it an excellent spot for the development of fishing activities.

CHAPTER 3

RESEARCH PROCEDURES

To begin the development of this work, the first step was to choose the subjects to be covered in the research, followed by a bibliographical survey of the themes applied to the study of environmental justice, artisanal fishing and SAO charters.

This was followed by a survey of works, articles, authors, reports and any other source of information that could contribute to the research by providing data. Following on from this procedure, reading was another process carried out in order to establish ownership of the information collected.

The theoretical framework could be organised on the basis of these surveys of works and information, serving as a basis for the construction of the methodological framework which determined the data collection procedures during the research.

The bibliographic review was carried out on academic production on environmental justice, SAO charters and artisanal fishermen, including all aspects related to the daily work of these social actors and fishing activities, with a view to their maintenance, namely: catching strategies (access to fishing sites, camps, gear, etc.), fishing calendar, social organisation and division of labour, among others.

There was also a survey of works on the relationship between oil activity and fishing, information on the oil project in Tramandaí and the collection of data on the oil spill accident in the region.

This was done in order to obtain knowledge through the theoretical framework for the development of the research. This made it possible to establish ownership of the concepts and then define the research premises.

The premises of the research in the preparation of the study establish the appropriation of knowledge, subsequently giving rise to the possibility of creating hypotheses and formulating a research question.

In order for the research question to be answered, the hypotheses need to be tested. This is why the development of this work was based on a theoretical and analytical process, based on four research instruments, which made it possible to obtain data and information during the application of the study. The first was field observation, carried out during the primary data collection campaigns in the socio-economic area as part of the project to draw up Oil Sensitivity Charts for the Pelotas Sedimentary Basin.

Field observation contributed to data collection and the acquisition of information. At this point it was possible to reconnoitre the field, helping the researcher to identify and obtain evidence regarding the aspects that guide the behaviour of individuals.

The second came from the notes in a field diary, in which all the procedures carried out during the research were written down, along with the main information arising from the facts and events that occurred during the development of the activities that took place.

The field diary was used as a report, a notebook in which all the observations were recorded and all the procedures for the activities carried out during the study were described. It also recorded the names of the interviewees and the entire guidance schedule for the activities carried out in the field.

The third source of data used qualitative social research as its main method, based on semi-structured interviews with key players representing some of the bodies and institutions involved in the oil spill accident in January 2012, which hit the beaches of Tramandaí and Imbé. An interview was also held with a representative of the MMA about the management tool SAO Letters.

The use of semi-structured interview techniques allowed for more in-depth coverage of certain subjects. This was due to the advantage in terms of time elasticity and in relation to the issues that arose during the interviewees' approach.

The fourth consisted of two previously scheduled workshops with groups of artisanal fishermen from Tramandaí and Imbé, differentiating these groups according to their fishing environment, which could be marine or lagoon/estuarine. The main aim of the workshops was to identify the main socio-environmental problems caused by the oil spill accident and relate them to the principles of environmental justice.

To analyse the data, the results obtained from the interviews and workshops were triangulated and analysed with the elements of Environmental Justice and the SAO Charters.

> [...] the basic objective of the triangulation technique is to cover the maximum breadth in the description, explanation and understanding of the focus under study. It is based on the principles that it is impossible to conceive of the existence of a social phenomenon in isolation, without historical roots, without cultural meanings and without close and essential links to a social macro-reality (TRIVINOS, 2008, p. 138).

3.1. DESCRIPTION OF ACTIVITIES

3.1.1 FIELD OBSERVATION and FIELD DIARY

For the field observation procedure, all the information regarding the study environment to be visited was used. The objectives motivating the work were listed and, based on this, the field was recognised in accordance with the information collected in the bibliographical survey.

During this phase, a visit was made from 10 to 13 June 2013 to reconnoitre the field, see the research scenario, identify the communities and groups in the region and then identify the social actors who would contribute to providing the information to be collected during the study.

This moment was marked by the monitoring of data collection by the socio-economic component team for the Environmental Sensitivity Charts to Oil Spills in the Pelotas Maritime Basin project, carried out by FURG/UFRGS, with the participation of the author of this work as part of the team that captured the information.

The SAO Charts project helped the author to familiarise himself with the development of the SAO Charts management tool, making it possible to carry out activities during the course of the research that allowed for an analysis within the scope of the study.

The data obtained from the field observation carried out in the SAO Charts project was appropriated in accordance with the methodology suggested by the proposal drawn up by the Ministry of the Environment in the document "Specifications and technical standards for drawing up environmental sensitivity charts for oil spills", MMA (2004).

The information collected during the research was all recorded in a field diary, which contained all the data relating to the observations visualised, along with all the procedures taken during the study.

In the field diary it was possible to record all the observations made during the collection of socio-economic data for the SAO Charter, the interviews and the workshops.

The field diary was used as a means of recording the information and data collected, and was organised according to the activities carried out and procedures taken during the research.

3.1.2. INTERVIEWS

Semi-structured scripts were drawn up for the interviews and an audio recorder and field diary were used. The scripts were open to interventions or changes of direction for issues that might arise during the course of the interviews.

The purpose of the interviews was to evaluate the construction of the management tool Cartas SAO and to establish knowledge about the oil spill accident in the municipalities of Tramandaí and Imbé.

The scripts were based on the specific objectives determined by the research. This was done taking into account previous knowledge of the region, the existence of oil activity, the formulation of an operational chart of the area in the aspects of the SAO Charts, the occurrence of an oil spill accident and the presence of artisanal fishermen in the region.

Once the scripts had been created and the interviewees had been identified, the interviews were scheduled with the participation of some of the institutions and public bodies in the municipalities involved in the accident in the region.

The first interview took place on the afternoon of 11 June 2013, during the first visit made to the municipalities. There was a double participation, with the presence of members of the Imbé Department of the Environment and Fisheries.

The second interview took place on the morning of 12 June 2013 and was conducted with an employee of the Federal University of Rio Grande do Sul/CECLIMAR.

The third interview took place on the afternoon of 12 June 2013 with a member of the Tramandaí municipality's Environment Department.

The fourth interview took place with a member of the Tramandaí Fishermen's Union on the morning of 13 June 2013.

The fifth interview took place on the afternoon of 13 June 2013 at the Tramandaí Department of Fisheries and Agriculture, where the interviewee and other officials from the department were present. The interviewees did not authorise the recording of the interview, but they allowed the questions to be asked and provided the answers and notes of the information in the field diary.

The sixth interview took place during the second field visit on the afternoon of 14

February 2014, with a representative of the EMATER office in Imbé and also a participant in the North Coast Fisheries Forum.

In order to better understand the SAO Charts, the document *"Specifications and technical standards for preparing environmental sensitivity charts for oil spills"*, which guides this instrument of Petroleum Environmental Management, was analysed and a focused interview (Minayo, 2013) was carried out with an MMA manager, who had actively participated in its conception since 2001. This interview, which took place on 23/05/2014, aimed to clarify any doubts about the procedures, guidelines and consolidation of the instrument that had not been properly resolved when reading it, thus forming the framework submitted for analysis.

In a similar way to the other studies carried out in the context of Oil Environmental Management, the decision was made to keep the interviewees anonymous. This option is intended to allow the public manager to give his opinion on the issues dealt with without fear of retaliation, given that they are public figures.

In total, eight people were interviewed in seven meetings which led to seven interviews, one of which was conducted in pairs, as can be seen in the case of the Imbé Environment and Fisheries Department (Chart 1).

All the interviews were recorded, with the exception of two, such as the Tramandaí Fishing and Agriculture Department and the MMA manager. These were recorded in a field diary and the others were digitally transcribed.

Table 1. Organisation of the interviews.

Interview	Date	Shift	Location
	11/06/2013	Afternoon	Imbé Environment and Fisheries Department
25	12/06/2013	Morning	CECLIMAR-UFRGS
35	12/06/2013	Afternoon	Tramandaí Environment Department
45	13/06/2013	Morning	Trade Union Tramandaí fishermen
55	13/06/2013	Afternoon	Tramandaí Fisheries Department
65	14/02/2014	Afternoon	Office of the EMATER of Imbé
75	23/05/2014	-	MMA

3.1.3 WORKSHOPS

The workshops were held in conjunction with the second visit to the field, from 12 to 15 February 2014. At that time, the research activities were associated with data collection for the project "Analysing the Production Chain of Fish from Artisanal Fishing and/or Family Aquaculture in the State of Rio Grande do Sul", which is the result of an agreement between the University and the Rio Grande do Sul Department of Rural Development, Fisheries and Cooperatives (SDR).

The production chain project was looking for information on fishing activities in the region comprising the municipalities of Tramandaí and Imbé. In order to acquire more in-depth knowledge, the researcher was accompanied by the activities carried out with these researchers during field visits.

The information collected for this study was based on the interests of the research, seeking to meet the objectives of the work, selecting only the subjects that would be of interest to the researcher in analysing the SAO Charter instrument, together with the precepts of environmental justice.

The visits for the workshops were scheduled with the help of an employee from the EMATER office in Imbé. Soon after contact was made, the groups were delimited, and it was made clear that, in order to choose the participants, we would have to take into account indications from the community, from EMATER, which has extensive knowledge of the region, and from representatives of the Fishermen's Forum of the North Coast of Rio Grande do Sul, as a strategy for acquiring people who would provide input for the qualitative component of the research.

Two workshops were held. One took place in the morning in a house in the Mariluz neighbourhood, located in the municipality of Imbé, and was attended by twelve artisanal fishermen, including men and women who practice the activity. They work in the marine and lagoon/estuarine areas of the Tramandaí and Imbé regions.

The other workshop was held in the afternoon, with a very small group of just three fishermen, located in the centre of the municipality of Imbé, where they also fished in both the marine and estuarine areas.

We drew up an attendance list for the two workshops so that we could keep track of the names of the people present and keep track of the identification of the participants. We took photographs to record the event and the final products resulting from the activity.

The purpose of the workshop was to establish with fishermen, using participatory methodologies, a characterisation of fishing in relation to oil activity in the region, involving the participation of artisanal fishermen from Tramandaí and Imbé.

One of the activities carried out, of relevance to the research, was the creation of visual maps, with groups divided into fishermen from the marine area and the lagoon/estuarine region. Each group filled in the map with information on the fishing gear used, fishing areas, gear used, the main species caught and finally they were asked about the area affected by the oil spill accident on 26 January 2012.

Another activity involved drawing up a problem matrix. Based on the reports during the workshops, the fishermen were asked about the oil industry's activities in the municipality of Tramandaí, the problems caused by the development in fishing activity, the configuration of the accident and its main problems in relation to fishing.

All the data from the workshops was organised in the field diary, making it easier to see the observations made during the procedures applied. Each of the workshops generated a report, according to the project team's procedure.

The results made it possible to draw up four mind maps and two matrices with the fishermen, which were then systematised to generate a single map and matrix, respecting the information collected and the methodologies applied in the workshops held.

CHAPTER 4

RESULTS AND DISCUSSION

4.1.OIL SPILL SENSITIVITY CHARTS AS A TOOL FOR ENVIRONMENTAL EMERGENCY ACTION

The application of policies for the prevention and control of oil incidents is part of the broad context of environmental management in Brazil, governed by the competences and instruments determined by Law 9.966/2000 (BRASIL, 2000), which acts as a complement to the International Convention for the Prevention of Pollution from Ships MARPOL 73/78, signed by Brazil and promulgated by Decree 2.508/98 (BRASIL, 1998).

In the documents formulated in this convention, it is also possible to find in its Annexes, texts that deal with regulations for the prevention of oil pollution, which can be observed in the following aspects: issuing certificates and surveys; requirements relating to the construction of vessels, equipment; control of the operational discharge of oil; aspects of the prevention of pollution resulting from an oil pollution incident (MARPOL 73/78, 2012).

In the same way, the Oil Law also refers to the International Convention on Oil Pollution Preparedness, Response and Co-operation - OPRC/90. Promulgated in Brazil by Decree No. 2.870/98, in addition to assigning obligations to the Parties involved (signatory States to the Convention), the OPRC/90 provides for the development of a series of instruments for the prevention and control of oil incidents. These include: Emergency Plans for Oil Pollution; the National Contingency Plan; and International Co-operation between Member States (BRASIL, 1998).

The proposal to map areas sensitive to oil spills is in line with Law 9966/2000, known as the Oil Law. According to the Ministry of the Environment - MMA - which is legally responsible for the mapping, it was necessary to understand the socio-environmental characteristics of the coastal zone in order to subsidise the environmental management of oil activities, in the exploration, production and transport stages, from three perspectives.

The first aims to guide planning decisions on the presence or absence of oil activity in a given area, based on the socio-environmental assets present in the Sedimentary Basin, i.e. planning on a scale appropriate to the regional level. From the second perspective, the information mapped subsidises environmental licensing of the set of socio-environmental assets in the area where a given oil activity wants to be installed, thus subsidising decision-making on whether or not to authorise this activity. This is mapping on an intermediate scale. The third, which is the noblest objective of mapping, is to subsidise environmental emergency action. Thus, based on the knowledge of the coast's environmental attributes, the decision on emergency action is made.

In this sense, the SAO Letters make up the National Oil System (SisNÓleo) and are key elements of the Individual and Area Emergency Plans and the National Contingency Plan, as well as other instruments provided for in the Oil Law that make up the environmental management of oil activities.

While debating the appropriate scope for mapping and investigating the international literature on the subject, between 2000 and 2004, the MMA identified that the methodology drawn up by the National Oceanic and Atmospheric Administration - NOAA in the design of *"Oil Spill Sensitivity Charts"* made a significant contribution to public policy objectives.

According to Lima et. al. (2008), because the coastal zone is susceptible to accidents involving oil spills, Petrobras developed a coastal mapping methodology that explains the different levels of sensitivity of the intertidal zone to oil spills. This was done in accordance with the parameters and methodologies presented in the NOAA sensitivity mapping system, with adaptations to Brazilian peculiarities.

Furthermore, at the time, other institutions were also involved in this debate and had been analysing and testing methodologies for mapping areas sensitive to oil spills, including the MMA itself, which carried out some pilot surveys, such as the Potiguar-Ceará Basin.

This methodology was later adapted by the Ministry of the Environment (MMA), taking advantage of the experiences of PETROBRAS's Research and Development Centre (CENPES), consolidated in the "Basic Manual for the Preparation of Sensitivity Charts in the PETROBRAS System", in the formulation of the Specifications and Standards Techniques for the Preparation of Environmental Sensitivity Charts for Oil Spills (LIMA et. al. 2008).

As a result of these various contributions, the document "Specifications and technical standards for drawing up environmental sensitivity charts for oil spills", published by the MMA in 2004, organises and consolidates this oil environmental management tool. In other words, the proposal to map sensitive areas as a procedure, as provided for in Law 9.966/2000, culminates in a set of articulated products at different scales that make up the public policy instrument called the SAO Charter.

From reading Gundlach & Hayes, (1978); Michel & Dahlin, (1993); Araújo & Silva, (2002); MMA, (2004) it is possible to infer that the SAO Charts aim to spatialise information on socio-economic, biological, geomorphological and emergency structure aspects in order to determine environmental sensitivity to accidents involving oil.

As described in the document itself, the specifications were drawn up in conjunction with the Brazilian Institute for the Environment and Renewable Natural Resources (IBAMA), which is directly responsible for environmental control and licensing of oil industry activities, and the National Petroleum Agency (ANP), the oil sector's regulatory body, and took Petrobras' experience into account, given that the company had accumulated experience with the NOAA methodology.

According to MMA (2004), the international use of sensitivity maps means that, in addition to their use in contingency planning and the implementation of response actions to oil spills, SAO Charts have great potential for application in environmental planning in the coastal and marine zone, reinforcing the political and administrative instruments of territorial planning (MMA, 2004).

The process of adapting the NOAA methodology and designing the specifications for the Brazilian reality took into account contributions from the Brazilian Navy and representatives of the oil industry. In December 2001, a workshop was held to validate the document in Rio de Janeiro, with representatives from the MMA, IBAMA, ANP, the Brazilian Navy, the scientific community, state environmental agencies, the oil industry and interested NGOs. This stage consists of the participation mechanism adopted to define the scope of the SAO Charter.

According to the MMA (2004), the Specifications were submitted to the National Cartography Commission (CONCAR), in accordance with the legislation in force, so that the SAO Charts (Figure 2) prepared according to this methodology could be considered official Brazilian cartographic documents, for mandatory use in contingency planning, general damage assessment and the implementation of response actions to oil pollution incidents in the coastal zone and maritime areas under national jurisdiction.

There are 14 Maritime Sedimentary Basins in Brazil, covering an area of 1,550,000 km^2, of which around half (770,000 $km^{2)}$ are in waters up to 400 metres deep and the other half (780,000 km2) are in deep to ultra-deep waters (between 400m and 3,000m), a fact that has shown that it is appropriate to adopt the Maritime Basins as the cartographic unit (MMA, 2004). Consequently, a mapping effort is made for each sedimentary basin, or more than one in the case of the smaller ones. In other words, CARTA SAO involves a set of products: a database, sensitivity maps and an atlas, for each maritime sedimentary basin.

In practice, mapping takes place through an inventory of socio-economic, biological and geomorphological information about the coastal and marine zone. This inventory, standardised by the MMA (2004), is carried out using past data and on-site collections that give rise to a database. The information considered most strategic is organised in icons and an index, called ISL, which will be detailed later, and specialised in maps and an Atlas. The latter, in addition to the specialised information, presents an analysis of the mapped Sedimentary Basin.

According to the manager interviewed, NOAA's contribution consists mainly of incorporating the environmental sensitivity of the coastline to oil, defined by the Coastal Sensitivity Index (ISL), and mapping emergency response structures, in addition to simply surveying the biological and socio-economic attributes of the area, which are present in other methodologies.

The ISL establishes levels of sensitivity of the coastline based on its geomorphological characteristics, considering the type of substrate, slope and exposure to tides. The set of structures present makes it possible to understand the response capacity established in a given area - either due to the existence of enterprises dealing with oil or due to local socio-economic conditions.

In this sense, the presence of piers, fishing boats and meeting spaces take on a greater significance: that they provide initial capacity and are linked to emergency action. Or, on the other hand, to make explicit the absence of any capacity for action in the event of an oil spill. Both are relevant elements for decision-making in the different contexts in which the SAO Charter is used.

The determination of the ISL can be seen in the table below (table 1). It is organised on the SAO Charts according to the type of coastal exposure, with each index represented by a colour and highlighted on the respective coastline corresponding to its exposure.

Table 1. Representation of the ISL on SAO charts.

COR	ÍNDICE	CÓDIGO R	CÓDIGO G	CÓDIGO B	TIPOS DE COSTA
	ISL 1	119	38	105	▪ Costões rochosos lisos, de alta declividade, expostos ▪ Falésias em rochas sedimentares, expostas ▪ Estruturas artificiais lisas (paredões marítimos artificiais), expostas
	ISL 2	174	153	191	▪ Costões rochosos lisos, de declividade média a baixa, expostos ▪ Terraços ou substratos de declividade média, expostos (terraço ou plataforma de abrasão, terraço arenítico exumado bem consolidado, etc.)
	ISL 3	0	151	212	▪ Praias dissipativas de areia média a fina, expostas ▪ Faixas arenosas contíguas à praia, não vegetadas, sujeitas à ação de ressacas (restingas isoladas ou múltiplas, feixes alongados de restingas tipo "long beach") ▪ Escarpas e taludes íngremes (formações do grupo Barreiras e Tabuleiros Litorâneos), expostos ▪ Campos de dunas expostas
	ISL 4	146	209	241	▪ Praias de areia grossa ▪ Praias intermediárias de areia fina a média, expostas ▪ Praias de areia fina a média, abrigadas
	ISL 5	152	206	201	▪ Praias mistas de areia e cascalho, ou conchas e fragmentos de corais ▪ Terraço ou plataforma de abrasão de superfície irregular ou recoberta de vegetação ▪ Recifes areníticos em franja
	ISL 6	0	149	32	▪ Praias de cascalho (seixos e calhaus) ▪ Costa de detritos calcários ▪ Depósito de tálus ▪ Enrocamentos ("rip-rap", guia corrente, quebra-mar) expostos ▪ Plataforma ou terraço exumado recoberto por concreções lateríticas (disformes e porosas)
	ISL 7	214	186	0	▪ Planície de maré arenosa exposta ▪ Terraço de baixa-mar
	ISL 8	225	232	0	▪ Escarpa / encosta de rocha lisa, abrigada ▪ Escarpa / encosta de rocha não lisa, abrigada ▪ Escarpas e taludes íngremes de areia, abrigados ▪ Enrocamentos ("rip-rap" e outras estruturas artificiais não lisas) abrigados
	ISL 9	248	163	0	▪ Planície de maré arenosa / lamosa abrigada e outras áreas úmidas costeiras não vegetadas ▪ Terraço de baixa-mar lamoso abrigado ▪ Recifes areníticos servindo de suporte para colônias de corais
	ISL 10	214	0	24	▪ Deltas e barras de rio vegetadas ▪ Terraços alagadiços, banhados, brejos, margens de rios e lagoas ▪ Brejo salobro ou de água salgada, com vegetação adaptada ao meio salobro ou salgado; apicum ▪ Marismas ▪ Manguezal (mangues frontais e mangues de estuários)

Source: MMA, 2004: p. 68.

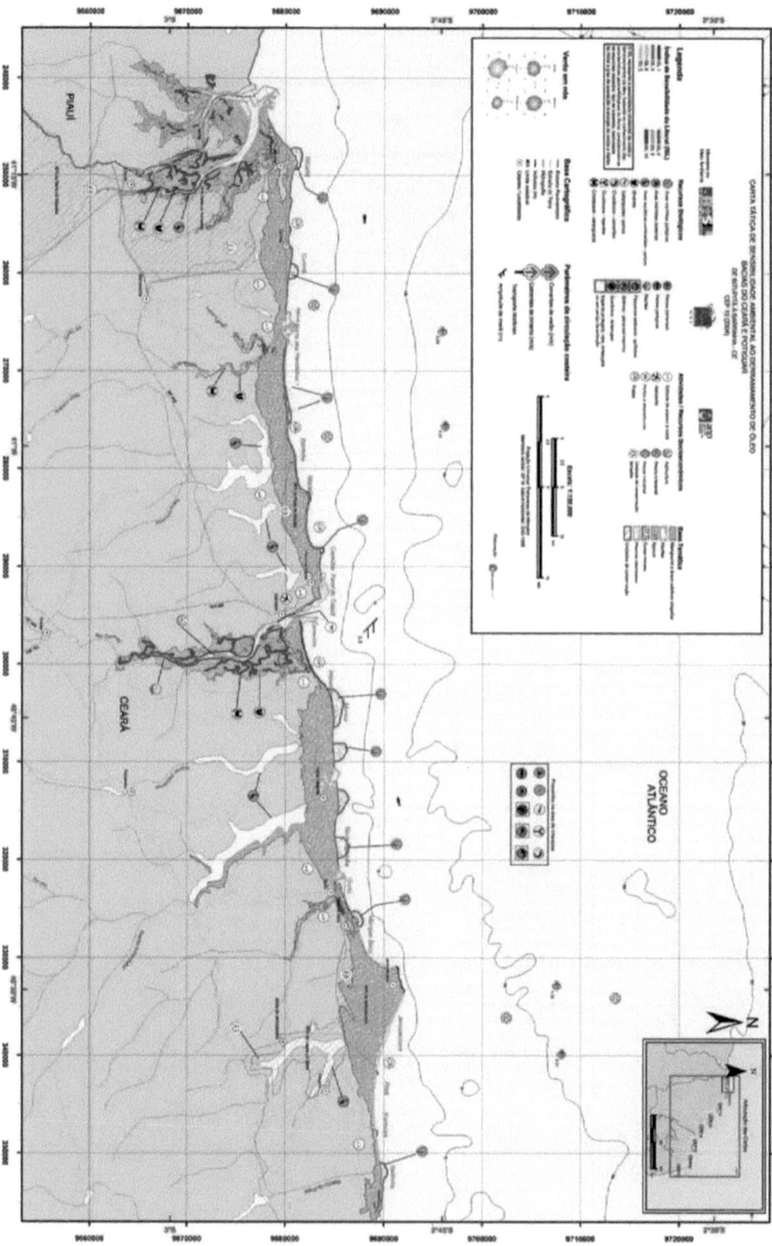

Figure 2 - Representation of a SAO chart.

Source: MMA.

All the mapping of areas sensitive to oil spills is carried out in accordance with a Cartographic Plan, taking sea basins as the cartographic unit and providing for mapping at three levels:

> *- Strategic (on a scale of around 1:500,000, covering the entire area of a given basin, or contiguous basins in the case of smaller basins);*
>
> *T Tactical (at a scale of 1:150,000, for the entire coastline of the mapped basin);*
>
> *Operational or detailed (at scales of 1:10,000 to 1:50,000, for high-risk/sensitive sites).*
>
> (MMA, 2004: p. 08).

With regard to the objectives pursued by the MMA and conceived through the technical specifications, the first objective of the SAO Charts is the Strategic Charts, which are mapped at a scale of 1:500,000, allowing for decisions at the macro level, which is the responsibility of the Federal Government. The second objective - to help with environmental licensing - unfolded in the Tactical Maps, which are mapped at a scale of 1:150,000. The third objective, to support emergency action, is possible from the Operational Maps, drawn up on a scale of 1:10:000 to 1:50,000 (MMA, 2004).

The stages that make up mapping and unfold in the SAO Charts are: i) collecting secondary and on-site data; ii) organising the information in a database; iii) drawing up the ISL and defining the attributes that will be included in the Charts; iv) making the maps, i.e. the SAO Charts at the three levels shown; v) drawing up an atlas.

In order to collect secondary data, it is necessary to carry out research into the region where the field activities will be carried out. Information is collected, such as information covering the entire socio-economic area, actions to respond to accidents, and the biological species of importance in the area.

Once this information has been gathered, the secondary data is organised into a table and then a field expedition takes place to observe the sites surveyed, capture new relevant data, validate the survey and mark the points on the GPS in order to identify the spatialisation of the sites visited.

Organising the information in the database currently requires the help of software, which is used to collect information and separate the categories of data. It contains all the information pertinent to the data captured in each resource.

To draw up the ISL, a field expedition is carried out, observing the coastline in order to identify the type of geomorphological composition of the coast, the type of substrate, the slope of the coastline and the degree of exposure to wave and tidal energy in each area, in order to finally define the sensitivity of the area, which will be classified according to the issues observed on a scale of 1 to 10.

To make the maps, i.e. the SAO Charts, the information collected in the field is used, where the most important data in each area is presented on the chart, represented in the form of icons differentiated according to the resource they represent, and spread out on the map according to their location. The cartographic standardisation suggested by the technical specifications and standards is used, and reference to the other information can be found by searching the database.The Atlas is the last procedure; it contains all the information on the regions surveyed. It is a means of publicising the work and the information collected. It contains the sensitivity maps and all the illustrative material relating to the mapped region.After all the steps have been taken, the product is: i) a database containing information on the geomorphological, biological and socio-economic attributes of the coastal and maritime zones and the emergency response structures present in the Sedimentary Basin, at different levels of detail; ii) maps, the number of which is delimited by Sedimentary Basin, at the three proposed levels: Strategic, Tactical and Operational; iii) an Atlas of Sensitivity to Oil in the Sedimentary Basin.

The table below (table 2) shows the three main types of information, which facilitate the observation of the resources present in the environment, helping to define the environmental sensitivity of the areas:

Table 2. Resources covered by the SAO Charters.

Environmental sensitivity of the coast to oil	Oil-sensitive biological resources in the charter area	Socio-economic activities
Defined by a Coastal Sensitivity Index (ISL), established on the basis of knowledge of the geomor- fological characteristics of the coast, taking into account the type of substrate, the slope of the coast and the	With information at species level and special attention to protected, rare, threatened or endangered species, and to places where concentrations or important stages of the species' life cycle occur, such as feeding areas,	That could be damaged by oil spills or affected by response actions, including recreational, leisure and holiday areas on the coast, fishing and mariculture areas, areas under special management (conservation

degree of exposure to wave and tidal energy. The ISL ranges from 1 to 10, from the least sensitive to the most sensitive.	breeding grounds, nurseries, nesting habitats and transit areas/migration routes.	units, historical or cultural sites), water intakes for industrial or energy plants, salt pans, ports and terminals, etc.

Source: MMA (2004).

A relevant aspect to consider from the interviewed manager is that when the technical specification was conceived, there was not enough information technology to make the information from the database dynamic, which made it static and the main product, the printed letters. Today, however, the database is the SAO Charter's most important product, as it gives managers access to information

It also cross-references these data, generating specific maps. He points out that since 2011 the MMA has made the databases of all the mappings carried out available to anyone on its website. Or it sends them digitally on request.

The Charts, which are also available digitally on the website, refer to the product generated and the decision-making based on the data initially standardised for their composition. However, it is the database that establishes the detail of the data and the manager's greater decision-making capacity.

The atlas, on the other hand, serves to publicise the instrument and is distributed to institutions working in the mapped sedimentary basin. It contains a chapter analysing the Basin, its socio-environmental attributes and aspects relevant to environmental planning.

According to the MMA (2004), in addition to the charts, the atlas should provide additional information that cannot be represented graphically, such as: (i) a description of the coastal habitats occurring in the region, with their respective ISL; (ii) information on predicting the behaviour and persistence of the oil spilled in said habitats; (iii) a list of the biological resources existing in the basin and data on ecological density/concentration, seasonality, special phases of the biological cycle and information on species protected by law, rare, threatened or in danger of extinction; (iv) complementary information on socio-economic activities that could be affected by oil spills in the basin, (v) data that could facilitate access to impacted areas for response operations (containment and clean-up/removal).

Finally, it can be seen that the mapping of each basin has been carried out primarily by universities working in the coastal zone and has contributed to the training of human resources. However, the surveys are strictly technical and do not include mechanisms for the social participation of social groups located in the coastal zone, especially those located in areas where activity is concentrated and which could be impacted by an accident involving an oil spill, the targets of the Operational Charts.

4.2.ENVIRONMENTAL JUSTICE AS AN ANALYTICAL TOOL

The theme of environmental justice indicates the need for the environment not only to be understood from the perspective of preservation, but also of distribution and justice. It thus represents the conceptual framework needed to bring together, in the same dynamic, popular struggles for social and human rights, collective quality of life and environmental sustainability (MOURA, 2010).

According to Bullard (1994), the concept of Environmental Justice refers to the fair treatment and full involvement of all social groups, regardless of their origin or income, in decisions about access, occupation and use of natural resources in their territories.

Acselrad (2004) says that environmental justice is "the pursuit of fair treatment and meaningful involvement of all people, regardless of race, colour, origin or income in the drafting, development, implementation and enforcement of environmental policies, laws and regulations".

Moura (2010) reports that the concept of environmental justice emerged in the United States, centred on the struggle waged by ethnic groups affected by environmental racism, because in 1987, a scientific report released by the Committee for Racial Justice of the United Church of Christ denounced the links between environmental degradation and racial discrimination.

The study presented statistical data to show that the installation of toxic waste dumps coincided with that of black, Hispanic and Asian communities. It was an issue associated with a historical milestone, strongly linked to the emergence of environmental justice movements in the US that linked the anti-racist struggle with the defence of the environment.

According to Bullard (2002), the actors who began to unite in this movement advocated politicising the issue of racism and environmental inequalities, denouncing the logic that they believed was "always in the backyard of the poor". Still according to Bullard (2002), it is in this context that the notions of environmental racism can be seen, a movement that goes hand in hand with the struggles for environmental justice, characterising the discrimination of social groups based on their ethnicity, skin colour, religion, practices and activities.

According to Moura (2010), twenty years later, the data released by the report presented by the Racial Justice Committee of the North American United Church of Christ was reviewed by experts in the field of environmental sociology, who refined the initial research by introducing new, more precise techniques, revealing that the study reinforced the original finding by demonstrating that, in the process of choosing sites for hazardous waste disposal, the most determining factor is the "race", ethnicity, of the inhabitants. What's worse, this relationship has become even stronger in recent years.

Acselrad (2002) reports that delegates to the First National Environmental Leadership Summit of Colour, which took place from 24 to 27 October 1991 in Washington (USA), drew up and embraced 17 principles of environmental justice. Since then, the principles have served as a reference document for the growing grassroots environmental justice movement.

In Brazil, environmental justice has a broader connotation than in the United States. Here, environmental justice expresses a movement to re-signify the environmental issue, appropriating the theme of the environment by socio-political dynamics traditionally involved with the construction of justice in a broad sense. (MOURA, 2010: p. 05).

According to Herculano (2002), environmental justice gained ground in Brazil through the Brazilian Environmental Justice Network (Rede Brasileira de Justiça Ambiental - RBJA), created in 2001 with the aim of combating environmental injustice in the country. It is an organisation made up of representatives of social movements, NGOs, trade unions and researchers from all over the country, whose aim is to encourage thinking and action that links environmental struggles with struggles for social justice.

Acselrad (2004) points out that after the creation of the Brazilian Environmental Justice Network, and many debates, a declaration was drawn up expanding the scope of the complaints beyond the issue of environmental racism in the allocation of toxic waste, which had founded the organisation born within the black movement in the United States. The definition of the category of struggle for "environmental justice" was then broadened to include the set of principles and practices that ensure:

a) That no social group, be it ethnic, racial or class, bears a disproportionate share of the negative environmental consequences of economic operations, policy decisions and federal, state and local programmes, as well as the absence or omission of such policies;
b) Fair and equitable direct and indirect access to the country's environmental

resources;

c) Broad access to relevant information on the use of environmental resources and the disposal of waste and the location of sources of environmental risks, as well as democratic and participatory processes in the definition of policies, plans, programmes and projects that concern them;

d) Encouraging the formation of collective subjects of rights, social movements and grassroots organisations to be protagonists in the construction of alternative development models that ensure the democratisation of access to environmental resources and the sustainability of their use.

(ACSELRAD, 2004: p. 13 - 20).

In our country, this issue is still quite recent and in its infancy, since environmental injustice is unfortunately still a feature of the dominant development model in Brazil, such as unemployment, social distancing and job insecurity. The majority of the Brazilian population is exposed to strong environmental risks, whether in their workplaces, their homes or the environment in which they move around. These environmental risks are due to hazardous substances, lack of basic sanitation, housing located on dangerous slopes and on the banks of watercourses, which are therefore subject to landslides and flooding. Many other homes are located near rubbish dumps, on gas pipelines or under electricity transmission lines (MOURA, 2010: p. 06).

Based on this, Acselrad (2010) shows that we can make an analysis in the sense of expanded and reinterpreted by the social actors in Brazil, where they are understood as subjects of resistance to the production of environmental inequalities: "the victims of contamination of spaces that are not directly productive - the surroundings of large risk-bearing enterprises and the outskirts of cities where environmentally undesirable facilities are located".

In this way, they make the connection between the generic discourse on the future and the concrete historical conditions through which the future is being defined in the present. This is the strategic junction between social justice and environmental protection: by affirming that, in order to stop the destructive pressure on everyone's environment, we must start by protecting the weakest (ACSELRAD, 2010: p. 12).

This raises the question of "who are the weakest?" In this sense, it is necessary to understand the concepts of risk and vulnerability of social groups.

4.3. THE VULNERABILITY OF SOCIAL GROUPS TO RISK AS A CENTRAL

ELEMENT OF ENVIRONMENTAL JUSTICE: THE CONTEXT OF ARTISANAL FISHERMEN IN ACCIDENTS INVOLVING OIL SPILLS.

According to Walter & Anello (2012), environmental vulnerability can be understood as a set of characteristics that generate a certain social group's pre-disposition to the impacts and risks of carrying out a polluting activity.

Although environmental vulnerability is a consequence of risk, it is a relational concept, i.e. it depends on the relationship between social actors and the environment in a given territory (ACSELRAD, 2006).

According to Nicolodi & Petermann (2010), the concept of risk is associated with an event that may or may not be realised. However, the existence of risk is only constituted when there is a valuation of some asset, material or immaterial, because there is no risk without the notion that something could be lost. Therefore, you can't think of risk without considering someone who is at risk, or in this case, society.

Environmental justice identifies unequal exposure to risk as the result of a logic that makes the accumulation of wealth happen on the basis of the environmental penalisation of the most dispossessed (ACSELRAD, 2010).

In the context of environmental justice and the analysis in question, such as the analysis of environmental justice in relation to oil spill accidents, the vulnerability to be observed in this study refers to that of social groups and their activities, which are exposed to risk situations. In the case of this research, artisanal fishing is analysed as the activity most vulnerable to this type of accident.

According to Walter & Anello (2012), we can say that the concept of vulnerability is linked to that of environmental justice, since the gigantic social injustices cover up and naturalise a set of situations characterised by the unequal distribution of power over the material basis of social life and development.

If we look at it from this perspective, the statement by Acselrad et al. (2004) makes sense: injustice and discrimination therefore appear in the elitist appropriation of territory and natural resources, in the concentration of benefits enjoyed from the environment and in the unequal exposure of the population to pollution and the environmental costs of development.

Walter & Anello (2012) highlight artisanal fishing as an activity of great importance that should be maintained in the country, given that it is considered a historical heritage (DIEGUES & ARRUDA, 2001) and contributes more than 50 per cent of fish production, which is mainly supplied to the domestic market.

Artisanal fishermen are those who, when catching and landing all kinds of aquatic species, work alone and/or use family or unsalaried labour, exploring ecological environments located close to the coast, because in general the boat and equipment used for this purpose have little autonomy (DIEGUES, 1973).

The definition, according to Law 11.959/2009, states that artisanal fishing is that which is practised directly by professional fishermen, independently or on a family basis, with their own means of production or through a partnership contract, landed, and using small boats (BRASIL, 2009).

In line with this definition, Walter & Anello (2012) emphasise that artisanal fishermen require an intact environment, along with access to both fishing areas and areas on land for their social reproduction. Oil activity, in its various stages, alters environmental integrity, as well as restricting fishermen's temporary or permanent access to traditional fishing grounds.

Agreeing with this factor, the authors comment on the existence of even more factors that can be considered as aggravating the condition of vulnerability of artisanal fishermen to oil activity, along with other developments installed in the coastal and marine zone.

The first factor is considered to be of a legal nature, established by Article 225 of the Brazilian Federal Constitution of 1988, explained by the fact that the coastal and maritime zone belongs to the patrimony of the Union, and that according to the National Coastal Management Plan Law (Law 7.661/88), the beach is established as a public good and of common use to the people.

With regard to coastal and marine areas as federal property, Walter & Anello (2012) state that these environments should serve the community and have social objectives, and the fact that they are federal property is a positive aspect of Brazilian legislation. However, by authorising a particular development, such as an oil platform or a port, there is private appropriation of that space by a particular economic agent and the exclusion of other economic activities, such as artisanal fishing.

They also claim that the more intense the oil activity, the greater the exclusion of other economic activities and of the social groups that depend on this territory for their social reproduction. However, the fact that artisanal fishermen have historically made use of common goods for their social reproduction, both on land and at sea, has not been enough to define legal and/or administrative instruments capable of compensating them or keeping them in their traditional spaces.

The second factor arises from the consequences caused by scientific studies and technical formulations in relation to fisheries management, most of which have historically focussed on the "resource" and the "catch" and not on the "territory".

According to Walter & Anello (2012), this approach results in the predominance of a common view that fishermen can carry out their activity in any part of the maritime space and that they have homogeneously distributed fishing resources. Thus, in the debates surrounding the impacts of oil activity on fishing, it is often argued that the impact is insignificant because the activity in question will take up a tiny amount of space when compared to the entire area of the oil basin.

Diegues & Arruda (2001) state that artisanal fishermen have a great deal of empirical knowledge about their fishing areas, such as determining which species exist, the best times to practise the activity and also the main problems and environmental changes to be suffered by the environment.

This consideration of fishing on the part of the oil industry is inadequate, as it disregards the fisherman's intense relationship with the environment, excluding from its considerations the dynamics of fishing, which according to Walter & Anello (2012) are characterised by: i) the presence of the resource targeted by the fishery in a given environment, which occurs in space and time; ii) the fishermen's knowledge of the occurrence of certain species in specific environments, at a given period of the year and/or day; iii) and the fishing equipment they have at their disposal, whose size of the boat, its form of propulsion and the fishing gear generate the skill needed to access the resource.

According to studies carried out by Diegues & Arruda (2001), traditional knowledge is extremely important. The relationships that exist between men and nature enable these individuals to observe the behaviour and cycles of environments and animals, acquiring knowledge of their surroundings empirically.

From this, Walter & Anello (2012) note that even the distribution of the fishing resource does not define the distribution of fisheries, but rather their relationship with the environment and the fishermen's ability to access it, which is a consequence of the fishermen's traditional knowledge combined with the technology and fishing equipment they have.

The third factor that makes fishermen more vulnerable to oil activity is the historical process of political disengagement from participating in decisions about their production process and the use of their territory, which has resulted in economic, social and cultural losses (WALTER & ANELLO, 2012).

Bearing this in mind, oil activity can have impacts on fishermen's vulnerability, which can lead to fishing being suspended, areas being restricted, risks of collisions with boats and oil spill accidents.

This shows that the appropriation of fishing areas by the industry can have an impact on fishermen's vulnerability.

Therefore, Walter & Anello (2012) state that the impacts can fall on fishing activities because they are concentrated in rural areas, have low incomes and live with insufficient public services or because they are concentrated on the outskirts of urban centres, whose infrastructure is even more precarious. These factors contribute even more to the vulnerability of this social group.

In these respects, we return to Brazil's fundamental social rights, which, following the exposed situation of vulnerability of artisanal fishing, also fall back on the lack of respect on the part of industries and public bodies for the conditions necessary for the development and recognition of fishing as a work activity.

4.4. OIL ACTIVITY IN TRAMANDAÍ AS A CASE STUDY

4.4.1 THE OIL BUSINESS

According to data provided by Tramandaí City Hall and organised by Cotrim (2008), oil activity in Tramandaí began in 1968, shortly after the municipality was founded in 1965. Oil activities began with the inauguration of the Almirante Soares Dutra Terminal (TEDUT), which began operating with the installation of a maritime platform, transport pipelines and an onshore terminal.

The offshore platform, made up of the MN 601 and MN 602 monobuoys owned by PETROBRÁS, operates with two oil transport pipelines (PETROBRÁS, 2006) that extend from the sea to the TEDUT onshore terminal located in the municipality of Tramandaí.

The area where the MN 601 and MN 602 Monobuoys maritime terminal is located (Figure 3), is 6 kilometres from the coast where the beach in the municipality of Tramandaí is located (PETROBRÁS, 2006).

The Tramandaí Monobay does not carry out extraction and production, it only operates as a receiver, helping to transport oil cargoes brought in by tankers from various regions.

The system for transporting the oil that is unloaded at Monoboia is based on the existence of pipelines, which connect the marine operating area to an onshore terminal for depositing the product.

These pipelines cross the sea until they reach the land area of the coast, where

Tramandaí beach is located, where they are visually exposed and then buried again until they reach the PETROBRÁS land terminal.

Figure 3. Tramandaí monobay - TEDUT.

Source: O Globo (FEPAM press release).

In the Tramandaí beach area, the pipelines occupy a space near the Brazilian Navy lighthouse and the sport fishing platform, which is a tourist site in the municipality, running through the city along some streets, avenues and neighbourhoods (Figure 4) until they reach the RS 030 highway, where they skirt an area near the Tramandaí estuary, finally arriving at the TEDUT.

Figure 4. Signposting of the area where the pipelines pass through the city of Tramandaí.

At the TEDUT terminal (Figure 5), oil is stored and distributed by pipelines to the Alberto Pasqualini Refinery - REFAP, located in the municipality of Canoas - RS.

Figure 5. TEDUT terminal in Tramandaí.

4.4.2 ARTISANAL FISHING

According to Cotrim & Miguel (2007), there are a large number of families in the coastal plain region of RS, where the municipalities of Tramandaí and Imbé are located, who derive their livelihood from fishing.

According to the Ministry of Fisheries and Aquaculture, through the General Fishing Register (RGP), in 2011 there were around 869 artisanal fishermen in Tramandaí and 245 in Imbé .[1]

The coastal plain region where these two municipalities are located has strategic geographical conditions for the development of fishing, with a significant water supply for the activity, due to the presence of the sea and the Tramandaí river, with the existence of several lagoons and the formation of an estuary.

The Tramandaí lagoon is shallow, with sandy beaches to the south, marshes on its eastern shore and sandbank vegetation to the south-west. The municipality also has the

[1] Information obtained by the project "Analysing Fish Production Chains from Artisanal Fishing and/or Family Aquaculture" under the responsibility of LabGERCO through the Access to Information Act. Consultation carried out in March 2012.

Armazém and Custódia lagoons, all of which have estuarine characteristics (SARAIVA & PUPER, 1985).

These environmental conditions facilitate the appearance of fish in the region. According to Macedo et al. (2000), estuaries, which represent places of transition between continental and marine waters, known for their high fertility and as excellent producers and exporters of organic matter, function as true natural nurseries for many marine, estuarine and freshwater species, and are vital to the sustainability of the riverside populations that make use of their natural resources.

According to the workshops held with artisanal fishermen in the region, fishing is divided into marine and estuarine. The use of fishing gear and techniques differ according to the environment.

Cotrim & Miguel (2007) report that the process of specialisation in fishing has intensified in Tramandaí. According to them, there are various combinations between the social relations of the group of artisanal fishermen and the environment.

These interactions generated six fishing production systems within two zones, which the author calls agroecological. In the agro-ecological zone of the sea, the production systems were recognised as being cable fishing, tarrafa fishing and dinghy fishing. In the estuarine agro-ecological zone, production systems were identified for fishing with small aeroplanes, tarrafa shrimp and trade (COTRIM & MIGUEL 2007).

During the workshops, the artisanal fishermen reported that the main fishing gear used to fish from the beach is the tarrafa and cable fishing. Fishing further from the coast is carried out using boats, and less frequently than the others.

Cotrim & Miguel (2007) report that the harsh characteristics of the region's sea, where there are strong currents, did not make it possible to develop deep-sea fishing, leading artisanal fishermen to adapt and specialise. As a result, some fishermen began to explore the coastal lagoons, others opted to explore the estuary of the Tramandaí lagoon and many spread out along the long stretch of beach along the coast.

According to the fishermen, this is due to the limited resources available for this type of fishing, as well as the difficulties in accessing the regions.

The artisanal fishermen report that the main fishing gear used in the estuarine area is the tarrafa and nets attached to fishing poles. Fishing takes place most frequently in the

Custódia, Armazém and Tramandaí lagoons (Figure 6).

According to them, tarrafa fishing is done with a circular net that has a rope attached to it, where the fisherman throws the net into the water and pulls it through the rope, collecting the catch.

They report that fishing in the beacons takes place with nets attached to stakes set in the lagoon, where there is a rotation of use of these stakes up to the mouth of the Tramandaí river.

The literature shows that the main species of fish caught by fishermen are mullet (Mugil platanus) and catfish (Netuna barba). The main crustaceans that use the lagoon for their development are the pink shrimp (Farfantepenaeus paulensis) and the crab (Callinectes sapidus) (COTRIM & MIGUEL, 2007).

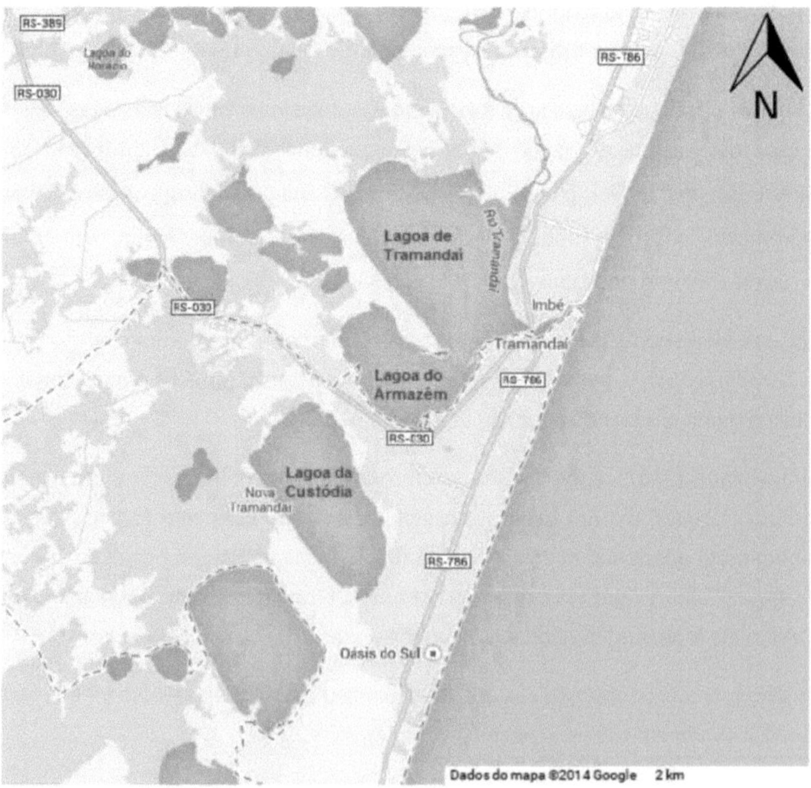

Figure 6 - Location of the lagoons within the Tramandaí River estuary.

Source: Google map data - 2014.

During the workshops, the fishermen provided data on the main species caught. In the estuary: mullet, catfish, corvina and sole. And on the beach: anchovy, catfish, white catfish, swordfish, angel dogfish, king fish, croaker, arrow bass, grouper, peba bass, sole, sardine, miraguaia, hammerhead shark, mullet.

With regard to the manner, area and period in which fishermen in the Tramandaí and Imbé region fish, there is state and local legislation on fishing.

guidelines to help them carry out their activities.

In the municipality of Imbé, there is municipal law no. 1328, which establishes the surfing and fishing areas along the beachfront throughout the year, advising beach users of the areas where they are allowed to practise their activities.

State decree no. 49.245 of 18 June 2012 provides for the demarcation of leisure, fishing and recreational areas in municipalities with a seafront, lake or river in Rio Grande do Sul. According to the decree, the aim is to guarantee the maintenance of activities using the spaces in a way that offers safety to all those who frequent these environments.

With regard to the sustainability of fishing, fishermen have been subject to rules on the criteria for fishing in the Tramandaí River Basin since the application of MMA Normative Instruction 17 of 2004. This normative instruction was replaced by IN 07/2013, with the aim of maintaining the activity and protecting species against predatory fishing measures.

Cotrim & Miguel (2007) report that a local fish marketing strategy was consolidated in Tramandaí based on the characteristics of low catches per production unit, individualised cold storage and a nearby consumer market. According to them, a local fish trade dynamic was generated, where fish was processed in an artisanal way and marketed directly to the consumer.

According to Perucchi et al. (2012), the North Coast Fishing Forum was set up in 2003 with the aim of holding an integrated dialogue on the problems surrounding artisanal fishing on the north coast of Rio Grande do Sul. Various players take part in the Forum, in addition to the associations, unions and fishermen's colonies, which organise the presence of fishermen at meetings. There is also governmental participation , with the presence of IBAMA, the Secretariat for Rural Development (SDR) through the Department of Fisheries, Aquaculture, Quilombolas and Indigenous Peoples (DEPAQUI), EMATER, the Ministry of

Fisheries and Aquaculture (MPA), universities and others.

Perucchi et al. (2012) also state that among the issues identified in the meetings with the fishermen, on which there is already a work of articulation and referral within the framework of the Forum of the North Coast of Rio Grande do Sul, the oil spill situation stands out.

Analysing the results of the workshops held with the region's fishermen, it is possible to identify from Walter & Anello (2012) that the characteristics of fishing vulnerability in the face of oil activity in Brazil are verified in the context of the fishermen of Tramandaí and Imbé (Chart 3).

Table 3. The vulnerability of fisheries and fishermen to oil activity.

0 what makes artisanal fishermen a social group vulnerable to the impacts of maritime oil activity? (Characteristics)	Why do the artisanal fishermen of Tramandaí and Imbé have such characteristics?	What impacts does oil activity have on the vulnerability of these fishermen?
Dependence on the environment to guarantee the conditions for their production and social reproduction.	1) They are a social group who depend on the intact coastal environment (lagoon, estuary and beach) for their artisanal fishing; 2) The fishermen had their activities suspended as a result of the oil spill.	Risk of accidents and consequent loss of environmental integrity (contamination, pollution, animal deaths, rejection of fish).
Lack of mobility due to: 1) Traditional knowledge of fishing areas and the dynamics of the environment on a local scale; 2) The use and development of equipment (gear, fishing techniques and boats) according to local characteristics.	1) They know the dynamics of the (estuarine) environment and the behaviour and population dynamics of the species; 2) They have equipment and fishing gear developed for use in the Tramandaí region.	Restriction of fishing areas due to the implementation of infrastructure, increased traffic of vessels supporting the industry and oil spill accidents.
1) Failure by public bodies to recognise the fundamental rights (health, education, work, housing, security, food safety) of traditional communities; 2)Lack of recognition of fishing areas as a means for fishermen to work.	There is no social security, social security or compensation for oil spill accidents.	The oil spill accident. Appropriation of fishing areas by industry.
Disrespect and disregard of fishermen's traditional knowledge in relation to scientific knowledge.	They make it impossible for fishermen to carry out their activities in certain fishing areas on the grounds that there are other places with better fish stocks and better fishing conditions.	Suspension of fishing, demarcation of areas, risks of collisions with boats and oil spill accidents.
Political alienation from participating in decisions about	Lack of political power that	Privatisation of areas with public

their production process and the use of their territory.	really represents the category.	use rights, restriction of areas for fishing.

According to the table, we can see that artisanal fishermen depend on the environment to carry out their activities, so they need to maintain the integrity of environmental quality.

With regard to the privatisation of areas of public use rights, this is characterised by the occupation of areas of common use owned by the Federal Government for the implementation and operation of the oil project, preventing public access to certain areas.

4.4.3. THE CONFIGURATION OF THE OIL SPILL ACCIDENT IN TRAMANDAÍ AND IMBÉ

a) ANALYSING THE INSTITUTIONS

The oil spill accident that occurred on 26 January 2012 off the coast of the municipality of Tramandaí and Imbé, according to the technical environmental report drawn up jointly by the Brazilian Navy, IBAMA and FEPAM (APPENDIX A), took place on 26 January 2012 and was quickly controlled and the environment recovered, with the case being resolved the following day.

According to the environmental technical report No. 01/2012/MB-IBAMA-FEPAM, all the appropriate measures were taken so that there was an agile and rapid resolution to the oil spill accident.

As for environmental recovery, the technical report shows that Transpetro/Petrobras, the company responsible for the leak at the MN-602 monobuoy at the Almirante Soares Dutra maritime terminal (TEDUT), provided all the repair assistance and environmental quality control, carrying out all the environmental analyses through third-party companies.

In the various interviews that took place regarding the accident, they said that it was a quick event and that it was resolved the next day. They gave the dimensions in terms of the size and location of the areas affected (Figure 7) and reported on the procedures carried out in response to the accident.

During the interviews, the interviewees were given a report (APPENDIX B) drawn up by the Imbé Department of the Environment and Fisheries, which dealt with the accident.

> "*According to technical report 003 of 2012, dated 1* February 2012, the survey carried out on 27 January 2012 found the presence of *small oil stains, mostly smaller than 1 cm²,* along the entire seafront of the municipality. *The Albatroz and Harmonia seaside resorts*

were the most representative in terms of visible presence, we didn't do a technical survey of oil presence and density, we did a visual survey of the area, saying here there is more, here there is less. So the Albatroz and Harmonia seaside resorts represented a greater accumulation of stains due to the tidal process. The fire brigade was contacted to prevent people from entering the sea, in order to safeguard people's physical integrity, since poisoning can occur through ingesting small clumps of oil. As well as skin and mucous membrane irritation from toxic and dangerous substances."

Talking to the Environment and Fisheries Department of the municipality of Imbé, it was possible to gather information on the impacts on the municipality's beachfront (Figure 8) and also on the municipality's image in the light of the accident.

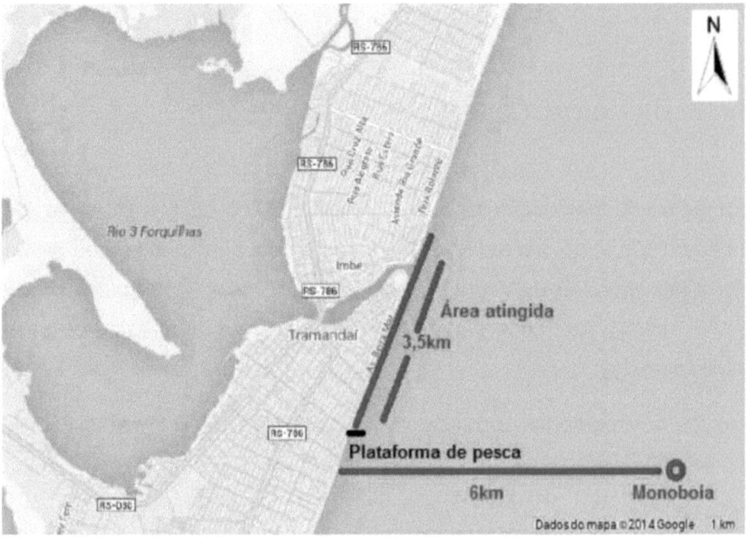

Figure 7. Location of the area affected by the accident.
Source: Google map data - 2014.

Figure 8. The beach contaminated by the oil slick.

Source: Correio do Povo newspaper online.

They said that the whole incident was very damaging for the municipality, as the accident occurred during the summer season, causing serious problems for traders, hotel owners and beach workers:

> *"It did a lot of damage to the visibility of the* municipality, because it was dirty, we had to do a general clean-up of the beach, but I believe that in general Imbé wasn't the worst hit. Tramandaí was much harder hit, the beach was completely covered in oil and it damaged an entire weekend for the kiosk owners, which represents thousands of reais in the purchase and turnover of these kiosk owners. In the summer they have two or three months, three months at the most, to work. This has had a huge impact on the *economy of the municipalities."*

The interviewees were asked about the situation of the region's artisanal fishermen, and they replied that the town hall played its part by warning the fishermen and warning them of the dangers posed by the accident.

They also emphasised the importance of fishing for municipalities, saying that he belongs to the Z-39 colony, that fishing is one of the main elements of the local economy and that because of this the town hall should pay close attention to and respect the activity of this socio-economic group that is of great importance to the town.

> *"The Department of Environment and Fisheries was created for the* government to get closer to these traditional communities here in the municipality, which have been occupying fishing grounds since the beginning of the occupation of the municipality in order to maintain the tradition of the fishermen and the families that support themselves on this

> product, which is fish. And yes, we realise that we're going to have to give full support to this class, which is very disadvantaged in terms of income and everything else. So we're going beyond that and we're developing other programmes to improve the conditions of *fish, fishing and fishing conditions."*

During another interview, he was told that the accident was brought under control quickly, that there was no environmental damage that could not be resolved and that the company responsible soon brought the accident under control, guaranteeing the quality of the environment.

In another interview, the Secretariat's participation in the action with Transpetro and the quality of the company's Emergency Plan were explained:

> "The *participation of the Environment Department was in support of some TRANSPETRO activity. TRANSPETRO has a whole contingency plan for these situations and when this happened, we monitored the scale of the event and then participated with TRANSPETRO in the remediation of the situation, especially the beach. Which is the direct effect it had on us, which is an apparent 3 or 4 kilometres of beach and beach land, which was the area most affected by the oil, and the operations, which fell precisely during the summer season, at the weekend, and had obvious economic repercussions, with complaints from all sorts of holidaymakers and traders who carry out some kind of activity on the beachfront. And the action was taken through the incidence of personnel and machinery at the time, with the aim of removing this oil from the sand on the beach as quickly as possible. We worked on this for several days. We carried out a full assessment together with TRANSPETRO, IBAMA, FEPAM and we monitored the whole event, but in any case that was the major visible effect. Other effects may have occurred in relation to the sea, which we obviously have no way of assessing. So FEPAM is in charge of assessing the repercussions. There is a process at the Public Prosecutor's Office that is assessing all these repercussions, including the company's liability, but so far this hasn't progressed."*

He pointed out that fishermen have a very hard life and that when the accident happened, the worst part was the week they were unable to fish from the beach for fear of contaminating the fish. He said that the accident only took a couple of days and that the main impact he saw was the oil stains all over the beach and some dirty birds.

When asked about the problems reflected in fishing, another interviewee said that they had only recommended that fishermen not fish during the period of the accident and that he didn't know if there had been any kind of compensation measure for them.

For another interviewee, the main impact was the presence of oil slicks on the beaches of Tramandaí and Imbé, which made it impossible to fish in these affected areas for a short period.

It is worth noting the similarity of the reports (Chart 4), both in relation to the emergency action and the beach clean-up, and in relation to the impacts on fishermen, which were considered to be restricted to the period when there was an oil slick in the water.

Table 4. Characterisation of the accident from the interviews.

Number of interviews	Duration of the accident	Areas affected	Main impacts	Main problems for fishermen
7	2 days	Tramandaí and Imbé beaches	Oil stains spread across the water and beach sand	Fishing suspended during the week of the accident

b) ANALYSIS OF ARTISANAL FISHERMEN

During the workshops, the artisanal fishermen recounted the oil spill accident, mentioning the areas affected by the oil slick and the main problems generated for fishing in the region (Chart 5).

According to the fishermen, the oil slick spread all along Tramandaí beach, from the sports fishing platform on the coast to the entrance to the estuary. On Imbé beach, the slick was concentrated only near the entrance to the estuary.

According to the artisanal fishermen, the oil spill that hit the beaches of Tramandaí and Imbé caused a lot of damage to fishing in the region.

They reported that the accident was brought under control quickly and that the next day the beach was clean, but they did face some problems in carrying out their activities.

Among the main problems listed in relation to this activity, the fishermen pointed out during the workshops that the suspension of fishing was the most prominent, caused by the accident.

According to the fishermen, the local environmental authorities, in partnership with the company responsible for the accident, recommended that the beach not be used and that no fishing activities be carried out in the area due to the pollution and contamination of the environment.

According to the fishermen, fishing was suspended for a week, preventing the reproduction of their activities and making it impossible to fish on the beach for consumption and commercialisation.

Another problem was the commercialisation of the fish. They reported that people didn't want to buy fish that had already been caught before the accident, because they were suspicious about the integrity of the product, linking the fish to contamination due to the accident in the region.

Fishermen also mentioned as a problem during the workshops that holidaymakers didn't come to the region as usual at that time and that people who were on holiday left earlier than planned, according to them because of the accident. Considering that fish is mainly sold locally in the summer, the absence of holidaymakers has a direct impact on fishing activity.

The main impacts of the accident on the environment pointed out during the workshops were the presence of oil on the beach sand, oil stains in the water, the bad odour left by the product, the presence of some oil-soiled birds and the death of some fish.

Table 5. Characterisation of the accident by the workshops.

Duration of the accident	Fishing suspension time	Areas affected	Main impacts	Main problems for fishermen
2 days	7 days	- On Tramandaí beach, from the sports fishing platform to the entrance of the	- Oil stains all over the water and beach sand, -Birds with oil stains,	- Suspension of fishing during the week of the accident, - Lack of commercialisation of fish,
		estuary, - On Imbé beach, near the entrance to the estuary.	- Bad odour, -Fish die-off.	- Lack of fishing for own consumption.

Perucchi et al. (2012) state that in 2012, following the oil spill in the Tramandaí sea area, the North Coast Fishermen's Forum sent a letter to the State Public Prosecutor's Office requesting that the competent authorities take measures to prevent this type of accident and asking for funds to compensate for the damage to fishing and tourism in the region. The only response to this request was a clarification from the IBAMA representative during a meeting, regarding the sanctions that the company responsible for the accident would receive, which would be: an administrative fine and an obligation to carry out socio-environmental repairs.

According to the artisanal fishermen, the importance of the beach for carrying out their work was excluded. They reported that the oil spill, despite damaging their fishing activities, did not result in any compensation for the damage caused, because according to them, no compensation was passed on to cover the losses and damage suffered.

4.4.4. THE CONFLICTS BETWEEN OIL ACTIVITY AND ARTISANAL FISHING BEYOND THE INCIDENT

a) URBAN SPRAWL

After the foundation of the municipality of Tramandaí with the inauguration of the Almirante Soares Dutra Terminal in 1968, oil flow activities began in the region.

As a result of these activities, many people have moved to the municipality in search of job opportunities in the oil industry.

As a result, the property sector grew to meet the demand of people arriving in the region.

According to Cotrim & Miguel (2007), from the point of view of official representations, during this phase local politicians shifted their interests from a development process linked to fishing to urban issues. It was during this period that the transition in the thinking of the local political group made up of mayors, councillors and deputies was consolidated.

As the city grew, certain areas became more valuable, the property market leveraged its development and took advantage not only of the oil business but also of tourism during the summer periods.

Cotrim & Miguel (2007) state that the dominant logic from this moment on was that fishing was an activity of the past and that the future would be given by tourism. In this way, all public efforts and initiatives favoured the urban, the holidaymaker and the construction industry. Significant proportions of these local representations were professionally linked to the property business. In this context, the fisherman gradually became a second-class actor and there was a long process of erosion of the category.

In this scenario, several fishing families moved to regions further away from the urban centre, some abandoning their homes and others offering their houses to real estate speculators, losing possession of their homes.

Vasconcelos et al. (2007) identify a crisis in the fishing sector, which is caused by external factors, such as rampant urbanisation in fishing territories, and internal factors, such as the lack of institutional organisation, both internal to the fishermen and the institutions involved in fishing.

Perucchi et al. (2012) state that fishermen recognise this problem and identify that the advance of urban occupation over their spaces harms their social reproduction and way of life. They criticise society's devaluation of professional artisanal fishermen and fisherwomen.

The expansion of construction was closely linked to the actions of local property developers. In the 1960s and 1970s, these players sought out land on the seafront for tourism projects and changed the profile of the region. Several communal sites that had previously been home to fishermen were formally legalised by notaries to become private property for sale to holidaymakers. The conflict over the legitimacy of land ownership in some areas is still a strong issue in the region today (COTRIM & MIGUEL, 2007).

According to Cotrim & Miguel (2007), many marshes that during the fishermen's village phase were flooded by the waters of the lagoons in the winter, serving as nurseries for fish and crustacean species, were landfilled and public buildings and housing estates built over them.

According to the artisanal fishermen, urban expansion due to population growth has brought with it various environmental problems in terms of development. In the case of the municipalities of Tramandaí and Imbé, we have the presentation of disordered growth that can be characterised as a lack of adequate infrastructure to meet the basic needs of the population, such as a lack of sewage systems and basic sanitation, negatively influencing fish stocks.

Cotrim & Miguel (2007) report that the work to fix the Tramandaí river bar and the construction of the concrete bridge between Tramandaí and Imbé made it difficult for schools of fish to enter the estuary to reproduce naturally. When building the dyke at the entrance to the bar, a shape was chosen that would prevent the strongest sea current coming from the south from having a direct effect, but this also hindered the natural entry of schools of fish such as catfish and mullet, which were trapped at the entrance to the bar and moved elsewhere.

If the shoals managed to enter the bar, only a few metres away they came up against the obstacle of the concrete bridge pillars, which ended up acting as a bulkhead, scaring the fish back into the sea (COTRIM & MIGUEL, 2007).

b) GENERATING LABOUR

According to Cotrim & Miguel (2007), before Tramandaí's political emancipation, the region's economy was basically geared towards artisanal fishing and commerce, with fishing being an activity of great importance due to the favourable natural conditions for its practice and also in terms of the number of families that used the activity for survival.

With the installation of the TEDUT terminal, a number of job opportunities linked to the oil business began to appear in the region. This led to a large number of people moving to the region in search of better job opportunities.

According to Cotrim & Miguel (2007), the arrival of Petrobrás in the 1960s, with a crude oil unloading point, was also important, because the company absorbed local labour, many of whom were fishermen, and also because of the work to fix the Tramandaí river bar, which changed the environmental dynamics of the estuary.

With the paving of roads giving access to the region and because it is located on the coast, during the summer holiday periods the town began to receive a large number of tourists, expanding the interests of the economy and changing the urban formation scenario of the town.

Cotrim & Miguel (2007) report that a change began to take place in the local fishing system, where at this stage the region's politicians shifted their interests from a development process linked to fishing to urban, industrial and tourist issues. From then on, the fisherman was put to one side and the concern for the municipality began to be the development of the economy based on trade and oil activities in the region.

Urban life has gradually changed, with bars and restaurants that only opened their doors in the summer now working day and night all year round. The city began to live two distinct lives: in the summer, it was filled with thousands of holidaymakers and offered various leisure options, and in the winter, it became a place for those people, mostly retirees, who were looking for quiet moments by the sea (SARAIVA & PUPER, 1985).

According to Saraiva & Puper (1985), the construction boom in this fishing system led some fishermen to leave their nets and tarrafas and become builders, bricklayers, carpenters and painters. With the presence of many hotels and holidaymakers' houses, fishermen's wives became cleaners in the summer.

According to Perucchi et al. (2012), there is also concern about the discontinuity of the category, which is a consequence of this crisis, in which the fisherman suffers various setbacks, discouraging the next generations, who end up giving up fishing and moving on to other spaces in society.

Nowadays, the scenario has changed in terms of absorption by the oil market in the region, with artisanal fishermen reporting that they have not been able to take advantage of the activity. According to them, this is due to a lack of qualifications to meet the demands of the new economic development model.

However, according to Cotrim & Miguel (2007), during this same period there was a significant increase in the number of fishermen who made their living from the lagoons and the sea.

Many families were attracted to Tramandaí to work in tourism during the summer months. These families continued to live in the city throughout the year and looked to fishing as a source of food and merchandise. In addition to Tramandaí's traditional fishing families, this phase saw the addition of people from the cities of the metropolitan region who fished to supplement their income or to help support their families. This situation increased the pressure on fishing stocks and led to a shortage of fish (COTRIM & MIGUEL, 2007).

c) VESSEL TRAFFIC

With the implementation of TEDUT's monobuoys, there has been an appearance and high flow of ships along the coast of the municipality of Tramandaí. These ships use the structure to handle their oil cargo.

With the large number of boats passing through the region, the artisanal fishermen reported in the workshops that they feel threatened because of the danger these vessels pose to their fishing activities.

The problems cited by the fishermen were the accumulation of waste dumped in the water, altered air quality, and the risk of accidents due to collisions with ships and oil spills.

Rosa (2003) states that the environmental impacts caused by ships, although less significant than those caused by other industries and modes of transport, are far from being considered irrelevant.

The sources of contamination, however, are not always as obvious as, for example, the use of anti-fouling paints. Some anti-fouling systems used on ships contain hazardous substances and pose a significant risk of toxicity and other chronic impacts on human life and marine organisms (ROSA, 2003).

According to Perucchi et al. (2012), in a study on artisanal fishing issues in relation to the Fishermen's Forum of the North Coast of Rio Grande do Sul, the environmental issues affecting the sea are pollution, predatory fishing and predatory exploitation of shellfish.

As for solid waste pollution, the fishermen point to the lack of awareness among people about how to dispose of it and the rejection of waste by large ships as the causes. As a consequence, there is damage to the fishing net, in which the waste is concentrated, and the fact that fish "escape" from the net, because according to the fishermen they see the waste (PERUCCHI et al., 2012).

Rosa (2003) states that it is not only oil that causes damage to the marine environment. Untreated sanitary waste from ships can pose a serious threat to the population due to the transmission of diseases and can also result in degradation of water quality.

With regard to the risk of accidents due to collisions between fishing vessels and ships, the artisanal fishermen report that they are instructed by the harbour master not to fish in the area close to the TEDUT monobay. According to them, this leads to damage to their fisheries, due to the restriction of the fishing area and exposure to the risk of accidents.

Although the fishing activities in Tramandaí and Imbé are not as characteristic of the high seas, with the exception of dinghy fishing, according to the workshops it was possible to observe that even so there is great concern about the dangers posed to the region and the fishing activity by the large influx of oil tankers.

According to the artisanal fishermen, these vessels jeopardise the quality of the environment and the fish, interfere with the life cycles of marine animals, drive fish schools away from the region and interfere with fishermen accessing the areas close to the installation of the monobuoys.

d) RESTRICTION OF FISHING AREAS

According to the Brazilian Navy's Hydrography and Navigation Department (DHN), the maritime oil terminal where the TEDUT monobuoys - MN 601 and MN 602 - are installed is located 3.2 miles south of the city of Tramandaí.

According to the artisanal fishermen, guidance from the harbour master's office is passed on, restricting the approach and release of fishing within a radius of 2 km from the development.

This recommendation, according to them, is due to the danger posed to fishing vessels by collisions with oil tankers and tugboats travelling through the region, or any implication of fishing problems with the installations of the offshore platform together with the oil transport pipelines.

Another problem that has already been addressed during the research in relation to the restriction of fishing areas is territorial conflicts, where according to Perucchi et al. (2012), the difficulty of access to fishing environments stands out, due to the great advance of real estate expansion on the shores of the lagoons on the North Coast.

The privatisation of these spaces prevents fishermen from being able to go to fishing spots or occupy the banks to rest or for protection when sailing in unsuitable conditions (PERUCCHI et al., 2012).

During the workshops with the artisanal fishermen, they reported and marked on mental maps the existence of support points for fishing, which could be characterised as deposits for materials and tools used in the activity.

These fishing support points are called sarilhos (figure 9), which are located on the banks of the Tramandaí lagoon estuary and play an important role in helping the region's fishing activity.

According to Perucchi et al. (2012), conflicts over territories are shaped by the different uses of the seafront, given that in some places fishing planning has not yet taken place, as required by state law 13.660/2011 (D.O.E., 2011), which stipulates that fishing, leisure or recreational areas must be demarcated in municipalities with a seafront, lake or river.

In meetings with the Fishermen's Forum and communities in the north coast region of Rio Grande do Sul, Perucchi et al. (2012) described that where these areas are demarcated, the participants noted that enforcement is stricter for fishermen, while other categories often go beyond the demarcated areas and also use the exclusive area for artisanal fishing, such as surfing, for example.

Figure 9. The Trouble

In the workshops, the fishermen reported that both the oil industry and all the other activities that have resulted from the development of the region have had an impact on fishing activities. Restrictions on the use of spaces considered important for fishing is a problem they pointed out, and according to them, this is due to the devaluation and lack of respect for fishing by society and public bodies.

4.4.5. THE INCLUSION OF FISHING AREAS AS AREAS OF ENVIRONMENTAL SENSITIVITY

The SAO chart, according to the "Specifications and Technical Standards" document for its preparation (MMA, 2004), is a management tool that assists in environmental emergency action in the event of an oil spill accident, presenting the information needed to draw up an accident containment plan, It is made up of environmental data, such as the geomorphological aspects that determine the environmental sensitivity index of the coast, the ISL in this case, information on socio-economic resources and data on biological species that are sensitive to the accident, organised in icons that strictly demonstrate their presence, without gradation.

In relation to the study area, incidents involving oil spills are recurrent, given the characteristics of oil transport in the area, a fact that required the preparation of Operational Charts. In this sense, it should be emphasised that the main task of these charters is oil spill contingency, which requires a detailed work schedule.

During the workshops it was possible to see that the characteristics of fishing vary according to the type of fishery and the target species, so a survey was carried out of the gear and techniques used in fishing. All the places where artisanal fishing takes place in the region were mapped, based on three guiding questions: What do they catch? Where do they fish? How do they do it?

The answers were superimposed on an illustrative map with a satellite image of the area for better visualisation by the fishermen. At this point they were able to highlight all the information about fishing activity on the map and provide the central elements that make up both the sensitivity of the area, based on their daily work, and the conflicts related to other socio-economic activities in the territory. Figure 10 summarises the characteristics of the fisheries in the environment.

When observing the two environments in which the study was carried out, estuarine and marine, it can be seen that there are major differences in the reality of fishing activity in these areas, delimiting the analyses set out below.

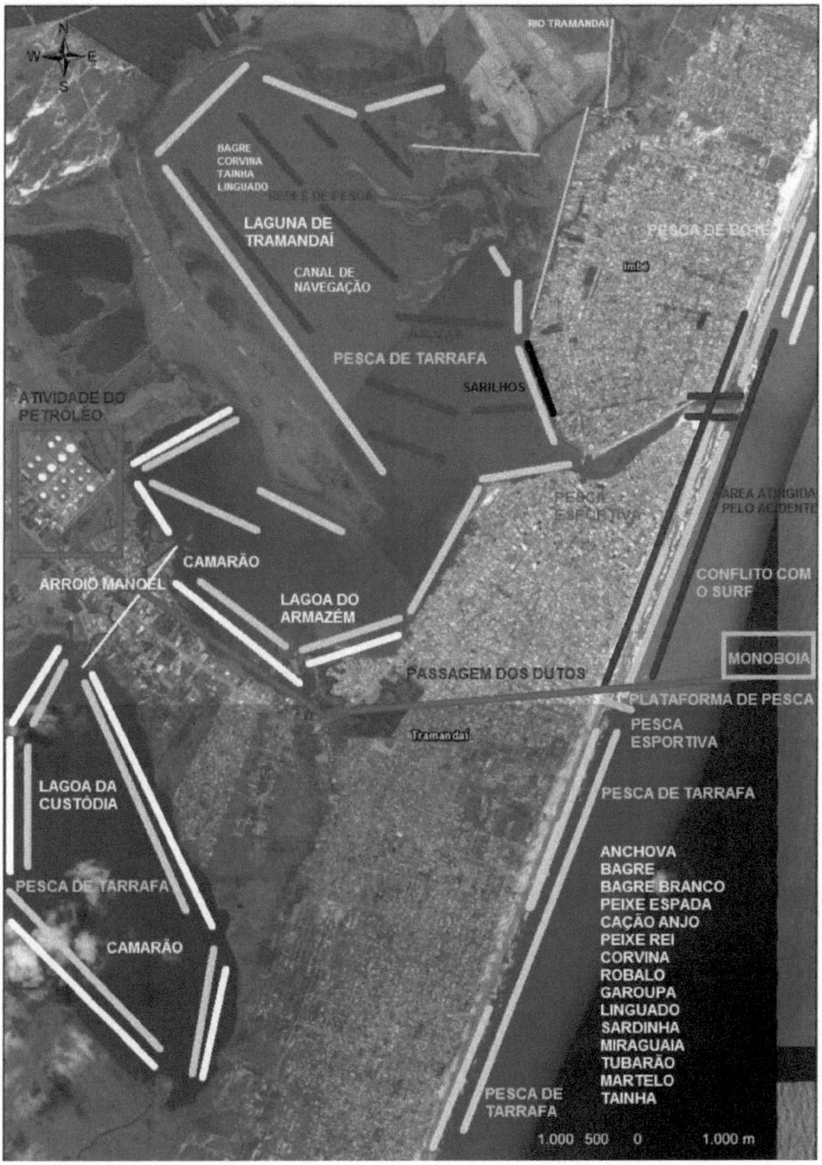

Figure 10. Map highlighted with the information provided by the fishermen.

4.4.6 THE ENVIRONMENTAL SENSITIVITY OF THE ESTUARINE ENVIRONMENT ACCORDING TO ARTISANAL FISHERMEN

During the workshops it emerged that there are fishermen who work in the estuarine environment, those who work in the marine environment and others who work in both.

However, estuarine fishing is the most frequent and most relevant to artisanal fishermen. This is due to the ease of access to the environment and also because it is less dangerous when compared to the physical conditions of the lagoon environment and the marine environment.

According to Cotrim & Miguel (2007), the coastal lagoons in this area are between one and two metres deep, which allows sunlight to penetrate the entire water column. This situation is favourable to the development of algae, which are the food base for fish and crustaceans, making this region a natural nursery for these species.

Cotrim & Miguel (2007) also state that these circumstances of shallow depth and estuarine environment have enabled the formation of marshes on the shores of the lagoons, which are environments rich in organic matter and play the role of holding excess water during the rainy season and releasing it during the dry season. In the marshes there is a complex plant formation of reeds and grasses that are essential in the food chain of lake species.

It was also observed in the workshops that these sets of conditions determine that these lagoons are considered to be prime locations for the development of typical estuary fauna, which has naturally created an environment rich in fishing resources.

During the workshops, the fishermen reported that the fish species caught in the Tramandaí lagoon are concentrated in certain areas. These are usually the areas where the fishing marks are located inside the lagoon and also in the channels and mouths of the rivers that flow into the estuarine complex.

Shrimp are also an important resource to be caught in estuarine fisheries. The main catch areas located by the fishermen at the time of the workshops are the areas covering Armazém lagoon, Custódia lagoon and the channel linking the two, called Arroio Manuel.

It should be noted that these two lagoons are very close to TRANSPETRO's land terminal, with the Armazém lagoon being the closest to the enterprise.

The Armazém lagoon is linked to the Tramandaí lagoon, which also has shrimp as a fishing resource. As these regions are interconnected, their waters have very similar characteristics, and they are all exposed to the dangers posed by oil activity.

4.4.7 THE ENVIRONMENTAL SENSITIVITY OF THE MARINE ENVIRONMENT ACCORDING TO ARTISANAL FISHERMEN

With regard to fishing in the maritime environment, it is necessary to emphasise the importance of the beachfront as a fishing spot, given that both the use of tongs and cable fishing are more frequent and the beach is the place where they are carried out. However, the use of the beach is of interest to various social actors who make use of this environment to carry out numerous activities.

These social groups on the beach compete for space with the fishermen, making it even more difficult for them to carry out their activities and giving rise to conflicts beyond the oil industry.

According to the artisanal fishermen, the oil activity is less damaging to fishing when compared to the disputes over territory that occur during the summer period with tourists and also over the surfing areas demarcated at the site. Thus, the vulnerability of fishing and fishermen to the oil project is great, but it is not isolated. It should be noted that other problems can accompany the presence of this activity in the region.

According to the workshops, it was possible to observe that these conflicts can occur in relation to the surfing activities practised by locals and tourists in the region. This led to the creation of Municipal Law 1328 of 12 September 2011, which establishes surfing and fishing areas, limiting the activities of fishermen and breaking a historical link between fishing in the region. For the fishermen, this law violates their rights, inhibiting and de-characterising the fishing activity carried out on the beach.

4.4.8 THE ENVIRONMENTAL SENSITIVITY OF ESTUARINE AND MARINE ENVIRONMENTS IN SAO CHARTS: AN ANALYSIS

Figure 11 shows the map of the Operational Chart for Tramandaí and Imbé drawn up in 2014 using the methodology of the MMA (2004), which is currently being revised. It can be seen that the lagoon area consists of an environment with a maximum ISL, i.e. 10, while the maritime area has a low ISL, with a value of 3. At first glance, the importance of the environments for fishermen seems coherent, given that the estuarine environment is more important than the maritime environment for their activities. However, both environments are characterised as single, homogeneous and do not take into account the relationship of artisanal fishermen within the environment. The icons show that fishing was not represented, despite other uses such as wind energy and recreation (beaches).

It should be noted that in the mapping that is currently being carried out, lagoon and marine artisanal fishing are described, but there is no distinction between the environments, nor within them, and it is only demarcated as "present".

On closer inspection, the methodology does not include any aspect that considers the presence of traditional populations as an element of environmental sensitivity, nor the use of specific areas necessary to maintain the social reproduction of these groups. Although a survey is made of existing socio-economic resources, historical aspects of relevant importance, archaeological sites and work activities, these do not result in sensitivity indices or criteria, in a similar way to the ISL. It is restricted only to the presence of icons, which determine the presence of the activity, in a similar way to all other socio-economic uses, such as the presence of industries, salt pans, recreation, etc.

This situation can be intriguing when we consider the precepts of Environmental Justice in the face of oil contamination in areas of significant expression and intense activity by traditional populations, ethnic groups, peasants and in the case of the subjects of this research: artisanal fishermen. All of these are also disregarded in terms of sensitivity in the international models of methodologies for SAO Charts.

Comparing the SAO Charts with the results of the workshops, incorporated in Figure 12, shows that, for the fishermen, the environment is not one. Lagoons and beaches are used in different ways within them, with symbolic representation and importance for the social reproduction of artisanal fishermen.

In both estuarine and marine environments, as a result of an oil spill, the homogeneity of the ISL throughout the same environment and the absence of other sensitivity criteria can result in an area relevant to artisanal fishing being chosen as a sacrifice, i.e. with a view to orientating the spot for later cleaning. This complicates the situation for fishermen who use this natural environment to guarantee their livelihoods.

The use of SAO Charts as a management tool to help choose less sensitive areas for the possible displacement of oil slicks is of fundamental importance in the decision-making process. It is with this information that the manager will apply common sense in relation to priorities for the protection of resources and certain areas of environmental preservation and conservation.

In the marine area, conflicts with oil, tourism and surfing activities are the main points that fishermen are informed about during the workshops. According to them, these activities compete directly with fishing, restricting their territory and jeopardising their practice. However, the choice of a fishing area as a sacrifice area, which is more dependent on environmental integrity, can result in greater damage to fishermen. Thus, management instruments must take into account the environmental conflicts present in the territory, since such conflicts often lead to practices of environmental injustice.

According to the results obtained from the workshops, when compared to the management instrument Cartas SAO, we can see that there is no mechanism to ensure the social reproduction of artisanal fishermen in the face of oil activity and that in the event of accidents, the protection of their activities is linked in the interim to the decision-making power attributed to the manager responsible for containing the disaster.

The information obtained from the workshops shows us that in the municipalities of Tramandaí and Imbé, there is intense competition for space, where fishermen are vulnerable to oil activities. In this case, therefore, greater care should be taken to ensure the significance of the fishing activity in its registration with the region's operational SAO Charter, or at least something that guarantees the social reproduction of this activity in the face of oil spill disasters.

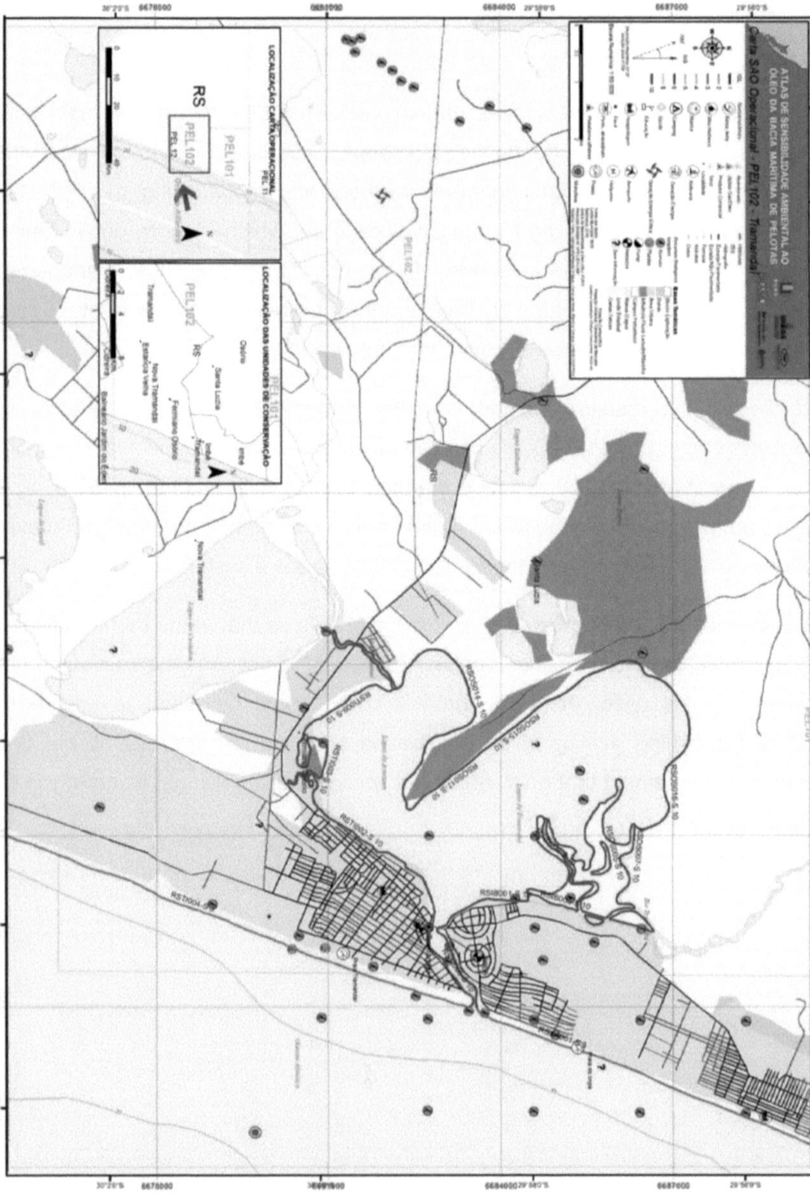

Figure 11. Operational SAO chart for Tramandaí.
Source: Pelotas Maritime Basin SAO Charts Project (Unpublished data).

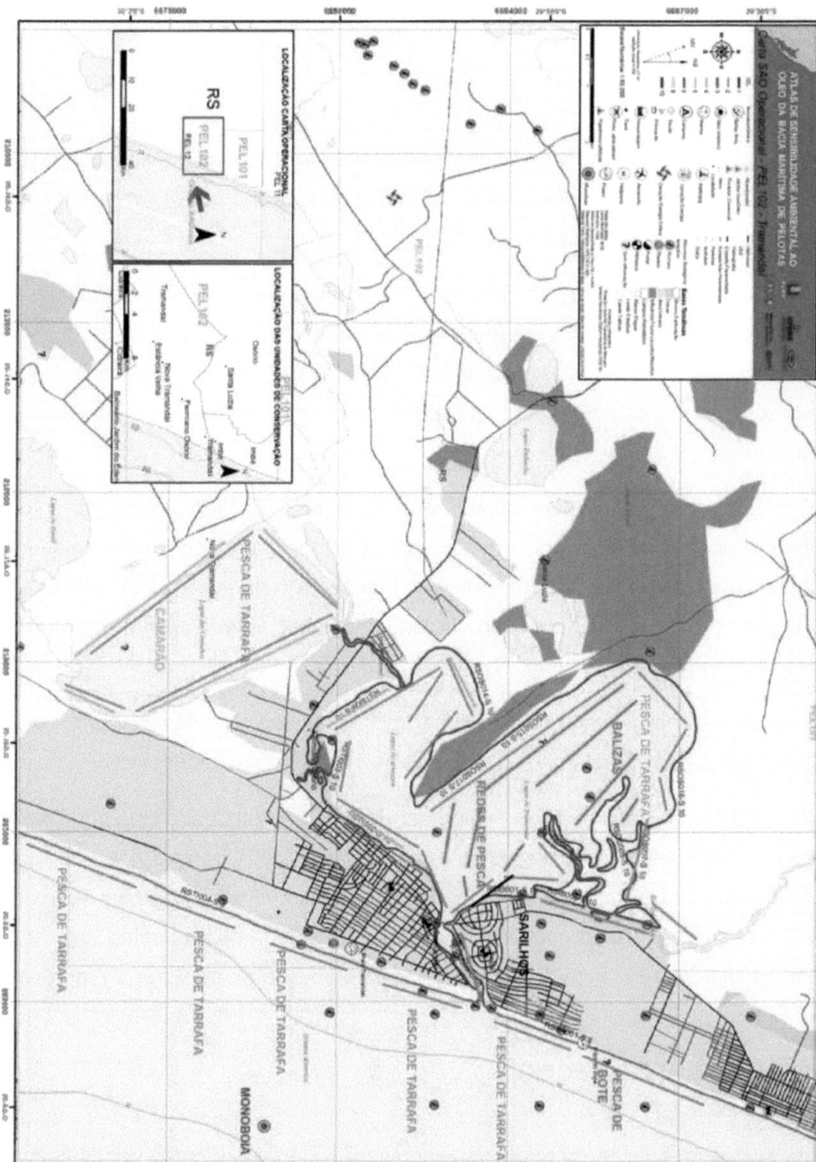

Figure 12. Representation of the fishing information collected in the workshops on the SÃO chart

CHAPTER 5

CONCLUSIONS

Analysing Environmental Justice in the context of Environmental Sensitivity Cards for Oil Spills is important in the development of this research with a view to improving environmental management instruments.

Because artisanal fishing occupies an important space within socio-economic activities and because it is a peculiar characteristic found in various parts of Brazil, it should be more respected by the authorities, society and public bodies. It should be recognised as vulnerable to the impacts caused by the oil sector or even by activities linked to economic development.

SAO charts, as an instrument used in the management of regions sensitive to oil spills, should be used in addition to their function as a tool to support the formulation of accident contingency plans. They should not only protect environments due to their intrinsic sensitivity exposed by the ISL, but also protect traditional activities and communities dependent on their environmental integrity.

The vulnerability of fishing should appear more exposed on SAO charts, making it protected from any impact by oil that could be directed at these areas of activity.

It would be fair and respectful of fishing activity, in accordance with the principles of Environmental Justice, if fishing appeared as sensitive in the SAO Charters in environments where the oil industry poses risks of environmental accidents. This would lead environmental managers to designate preferences for maintaining fishing, guaranteeing the social reproduction of the activity and less damage to the environments used by groups of fishermen.

BIBLIOGRAPHICAL REFERENCES

ACSELRAD, H (org.) Social conflict and the environment in the state of Rio de Janeiro. Rio de Janeiro: Relume - Dumará, 2004.

ACSELRAD, H (org.). Environmental Conflicts in Brazil. Rio de Janeiro: Relume - Dumará, 2004.

ACSELRAD, H. Environmentalisation of social struggles - The case of the environmental justice movement. Advanced Studies. N° 24 (68). 103 - 119p. 2010.

ACSELRAD, H. Environmental justice and the social construction of risk. Desenvolvimento e meio ambiente, n°5, p. 49 - 60. UFPR Publishing House. Jan/Jun - 2002.

ACSELRAD, H. Environmental vulnerability, processes and relationships. Communication to the II National Meeting of Producers and Users of Social, Economic and Territorial Information, FIBGE, Rio de Janeiro, 24/08/2006.

ADAME, Alcione and GAMBINI, Priscila Truviz. Measures for controlling and repairing accidents involving oil and oil derivatives provided for in national and international environmental protection legislation. Oil and gas law: environmental and international aspects / organisers Alcindo Gonçalves and Gilberto M. A. Rodrigues - Santos: Editora Universitária Leopoldianum: 2007, p. 177.

ARAÚJO, S. I.; SILVA, G. H.; MEUHE, D. Basic Manual for the Preparation of Environmental Sensitivity Maps to Oil Spills in the Petrobras System: Coastal and Estuarine Environments - Petrobras- RJ, 2002.

BAGGIO, R. C. Environmental justice between redistribution and recognition: the necessary democratisation of nature protection. 1ª. ed. Rio de Janeiro: Lumen Juris, 2014. v. 1. 301p.

BONI, V.; QUARESMA, S.J. Aprendendo a entrevistar: como fazer entrevistas em Ciências Sociais. Revista Eletrônica dos Pós-Graduação em Sociologia Política da UFSC, Vol. 2, n. 1, 2005. p. 68-8.

BRAZIL. 1988. Article 225. "Provides for the use of and access to the environment".

BRAZIL. 1988. Law No. 7.661 of 16 May 1988. Institutes the National Coastal Management Plan.

BRAZIL. 1998. Decree No. 2.508 of 04 March 1998. Promulgates the International Convention for the Prevention of Pollution from Ships, concluded in London on 2 November 1973, its Protocol, concluded in London on 17 February 1978, its 1984 Amendments and its Optional Annexes III, IV and V.

BRAZIL. 1998. Decree No. 2.870 of 10 December 1998. Promulgates the Convention International Convention on Oil Pollution Preparedness, Response and Cooperation, signed in London on 30 November 1990.

BRAZIL. 2000. Law No. 9.966, of 28/04/2000, which "provides for the prevention, control and monitoring of pollution caused by the discharge of oil and other harmful and dangerous substances into waters under national jurisdiction and makes other provisions".

BRAZIL. 2004. Ministry of the Environment. Specifications and Technical Standards for the Preparation of Environmental Sensitivity Maps for Oil Spills. p. 107.

BRAZIL. 2004. Ministry of the Environment. MMA Normative Instruction No. 17, of 17 October 2004. 6p.

BRAZIL. 2009. Law No. 11.959, of 29/06/2009, which "Provides for the National Policy for the Sustainable Development of Aquaculture and Fishing, and regulates fishing activities".

BRAZIL. 2013. Ministry of the Environment. MMA Normative Instruction No. 07/2013.

BRAZIL. Ministry of Planning, Budget and Management. Brazilian Institute of Geography and Statistics. Population Count. Available at: http://www.cidades.ibge.gov.br/xtras/home.php Accessed: 03/03/2014.

BULLARD, R. D. Dumping in Dixie: race, class and environmental quality. Boulder, Westview Press, 1990.

BULLARD, R. D. Environmental justice: strategies for building healthy and sustainable communities. In: II WORLD SOCIAL FORUM, Feb. 2002, Porto Alegre.

BULLARD, R. D. Grassroots flowering: The environmental justice movement comes of age. Amicus, 16 (Spring), p. 32 - 37, 1994.

TRAMANDAÍ RIVER BASIN COMMITTEE. Tramandaí river basin plan. Osório, RS, 2005. Available at: http://www.comitetramandai.com.br Accessed on: 03/03/2014.

COTRIM, D. S. Agroecologia, sustentabilidade e os pecadores artesanais: O caso de Tramandaí (RS). Dissertation for the Postgraduate Programme in Rural Development at the Federal University of Rio Grande do Sul. 2008.

COTRIM, D. S.; MIGUEL, L. de A. IN: Eisforia, year 5, volume 5, number 2. December 2007, PPG Agroecossistemas/ UFSC, Florianópolis. Pp. 136 - 160.

D.O.E. (2011) - Law No. 13.660, of 12 January 2011 - Amends Law No. 8.676, of 14 July 1988, which determines the mandatory demarcation of fishing, leisure or recreation areas in municipalities with a seafront, lake or river. Diário Oficial do Estado, 13.01.2011, section 010, p.1. Available at http://www.al.rs.gov.br/legis/M010/M0100099.ASP7HidTipo=TEXTO&Hid Tod asNormas=55733&hTexto=&Hid IDNorma=55733

DIEGUES, A. C. & ARRUDA, R. S. V. (eds.). Traditional knowledge and biodiversity in Brazil. Brasília: MMA; São Paulo: USP, 2001.

DIEGUES, A. C. Fishing and marqinalisation on the São Paulo coast. 1973. 187 f. Master's thesis - NUPAUB; CEMAR, University of São Paulo, São Paulo, 1973.

STATE OF RIO GRANDE DO SUL. 2012. Legislative Assembly. DECREE No. 49.245 OF 18 JUNE 2012 (published in DOE No. 117 of 19 June 2012).

Gundlach, E.R., Ruby C.H., Hayes M.O. and Blount A.E. (1978). The URQUIOLA oil spill, La Coruna, Spain: Impact and reaction on beaches and rocky coasts. Environmental Geology, Vol. 2, 3:131-143.

HERCULANO, S. Redesigning the debate on environmental justice: Theoretical production, a brief collection of cases and the creation of the Brazilian Environmental Justice Network. Desenvolvimento e Meio Ambiente. n. 5, p. 143 - 149. Editora UFPR. 2002.

HERCULANO, Selene & PACHECO, Tânia (Orgs.). Environmental Racism. Rio de Janeiro: FASE, 2006.

IMBÉ. Municipal Law No. 1328 of 12 September 2011. Demarcates surfing and fishing areas in the municipality of Imbé and makes other provisions.

Environmental Technical Report No. 01/2012/MB-IBAMA-FEPAM. Porto Alegre - RS. 03/04/2012, 9p.

LIMA, M. V.; DIAS-BRITO, D; MILANELLI, J. C. C. Revista Brasileira de Cartografia. N° 60/02. August 2008. 145 - 154p.

MACÊDO, S. J.; FLORES MONTES, M. J.; LINS, I. C. Chap. 2. Abiotic characteristics of the area. In: BARROS, H. M.; ESKINAZI-LEÇA, E.; MACÊDO, S. J.; LIMA, T. (Eds). Participatory Management of Estuaries and Mangroves. Ed. Universitária da UFPE, Recife, p. 7-25, 2000.

MARTINS, T. de A. P.; RIEDEL, S. P.; MILANELLI, C. C. J. Environmental oil sensitivity chart: Origin, evolution and trends. Revista Ciência e natura, Santa Maria - RS. v. 35 n. 2 December, 2013, p. 163-175.

MICHEL, J; DAHLIN, J. Guidelines for developing digital environmental sensitivity index atlas databases. Hazardous Materials Response and Assessment Division - NOAA, 1993.

MINAYO, Maria Cecília de Souza (2013). The Challenge of Knowledge: Qualitative Research in Health. São Paulo/SP: Hucitec Editora, 13th Edition, 407p.

MOURA, D. V. Justiça Ambiental: Um instrumento de Cidadania. Qualit@s Magazine. State University of Paraíba, Volume 09, n°01,2010. 10p.

Nicolodi, J. L.; Petermann, M. R. Climate change and the vulnerability of Brazil's coastal zone: Environmental, social and technological aspects. Revista de Gestão Costeira Integrada. (2010). n. 10, Vol. 2. p. 151 - 177.

PERUCCHI, L. C.; KUBO, R. R.; COELHO-DE-SOUZA, G. Articulating and forwarding artisanal fishing issues: An analysis of the North Coast Fisheries Forum of Rio Grande do Sul. Journal of Coastal Management. Vol. 12. N°4. Pág. 499 - 508, 2013.

ROSA, S. R. Z. Environmental management system applied to harbour tugs: The case of the port of Rio de Janeiro. Master's dissertation in environmental engineering - UERJ. 2003. p. 202.

SARAIVA & PUPER, S. Tramandaí land and people. Porto Alegre: Assessoria Gráfica e Editorial, 1985.127 p.

SEIFERT JR, C. A. Master's thesis in Coastal Management - FURG. The environmental governance of the prevention and control of oil incidents in maritime oil activities in Brazil. p. 277. Rio Grande - RS. 2013.

TRIVINOS, A. N. S. Introdução à pesquisa em ciências sociais: a pesquisa qualitativa em educação. São Paulo: Atlas, 2008.

WALTER, T.; ANELLO, L. de F. S. ENVIRONMENTAL EDUCATION AS A MITIGATING AND COMPENSATORY MEASURE: A reflection on the intrinsic concepts in relation to Oil and Gas Environmental Licensing with artisanal fishing as a context. Ambiente & Educação, vol. 17(1), 2012. 73 - 98p.

WIECZOREK, A. ; D, Dias-Brito ; MILANELI, J. C. Environmental Sensitivity Mapping Oil Spill os Cardoso Island Stat Park - SP. Ocean & Coastal Management, v. 50, p. 25/253-31,2007.

ANNEXES

APPENDIX - A

INTERVIEW SCRIPT

Gathering information on the oil spill accident that took place on 26 January 2012.

Name:

Occupation:

1- Find an account of the accident.
2- What was the organisation's involvement in the accident?
3- What are the main problems you see?
4- What area was affected?
5- How big was the accident? Size of the impacted area?
6- Who was most affected by the accident?
7- What is the main impact on those affected?
8- What's the situation like for fishing in the region?
9- What is the main impact on fishermen?

APPENDIX - B
WORKSHOP TO BUILD THE MAPS

Mind maps

*Separate by group: Lagoon or Marine.

1- Lagoon or sea fisherman?
2- What fishing gear do you use?
3- What kind of fish do you catch the most?
4- What is the best area to fish?
5- Where do you store fishing gear and instruments? (*There are small areas with marinas, ramps and covers to store boats).
6- Have you ever been prevented from fishing? What was it and why?
7- Has there ever been an environmental accident that damaged your fishing?
8- When the oil spill accident occurred, which area was affected?
9- Which individuals do you find most often in your fishing areas? (type of fish)
10- What state of conservation do you think your fishing environment is in?
11- When the oil spill occurred, which areas were affected by the oil slick?
12- Did this accident affect you in any way or did you continue to carry out your work?

APPENDIX - C

Photographs

Field observation with the Cartas SAO project group

Application of the workshops

Fishermen packing up their fishing nets

Fishing structures

MIX
Papier aus verantwortungsvollen Quellen
Paper from responsible sources
FSC® C105338

Printed by Books on Demand GmbH, Norderstedt / Germany